Mobile Secrets

Mobile Secrets

Youth, Intimacy, and the Politics of Pretense in Mozambique

JULIE SOLEIL ARCHAMBAULT

The University of Chicago Press
Chicago and London

The University of Chicago Press, Chicago 60637
The University of Chicago Press, Ltd., London

Published 2017
Printed in the United States of America

26 25 24 23 22 21 20 19 18 17 1 2 3 4 5

ISBN-13: 978-0-226-44743-8 (cloth)
ISBN-13: 978-0-226-44757-5 (paper)
ISBN-13: 978-0-226-44760-5 (e-book)
DOI: 10.7208/chicago/9780226447605.001.0001

Library of Congress Cataloging-in-Publication Data

Names: Archambault, Julie Soleil, author.
Title: Mobile secrets : youth, intimacy, and the politics of pretense in Mozambique / Julie Soleil Archambault.
Description: Chicago : The University of Chicago Press, 2017. | Includes bibliographical references and index.
Identifiers: LCCN 2016044856 | ISBN 9780226447438 (cloth : alk. paper) | ISBN 9780226447575 (pbk. : alk. paper) | ISBN 9780226447605 (e-book)
Subjects: LCSH: Youth—Mozambique—Inhambane—Social life and customs. | Cell phones—Social aspects—Mozambique—Inhambane. | Cell phone etiquette—Mozambique—Inhambane. | Courtship—Mozambique—Inhambane. | Communication and technology—Mozambique. | Inhambane (Mozambique)—Social life and customs.
Classification: LCC HQ799.8.M852 A73 2017 | DDC 305.23509679—dc23 LC record available at https://lccn.loc.gov/2016044856

♾ This paper meets the requirements of ANSI/NISO Z39.48-1992 (Permanence of Paper).

To Mia, my favorite person in the world

Contents

Figures

Acknowledgments

This book is the product of my love affair with Mozambique, which started in the mid-1990s when I first went there as part of a gap year spent traveling across southern Africa. It is the story of young people living in a suburb known as Liberdade—"freedom" in Portuguese—to whom I am forever grateful for sharing their lives and their colorful ways of seeing the world with me over the years. I feel incredibly privileged to have had the opportunity to conduct field research in such good company, not to mention in such an idyllic setting. No one said fieldwork had to be painful!

I am especially indebted to Kenneth Mangave, my research assistant and very dear friend, for proving such an able fieldworker and engaging interlocutor. I could not have written this book without his help. I am also grateful to Kenneth's family, which has become like a family to me, especially to Taninha, Felizarda, Mundo, Vivi, the late Papaito, and Tacita. Special thanks also go to Jhoker Macuacua and his brother Augusto, Jenny and Pajo, Omar Macuvele and Benvinda Zualo, Inocencio Vaz, Fakira, Bush, Lulu, Abibo, Lalo, Ketia and her brothers Mikas and Paito, Bud and Celinha, Belo, Balsa, Zumura, Bela, Inacio, Pascual, Emidio, Antonio, China, Almeida, Milas, Zuba, Genito and Julinha, Anaty Casamo, Virgilio Zandamela, Gina Bila and Innocent Chirua, Sarmento, Adolfo Gustavo, and the gang at the Kebrada in Maxixe. I am particularly indebted to Admiro, Mira, Neidi, Nono, and Denny for being the best neighbors and for their generous hospitality in more recent years, and to my friends at Pachiça, especially Zenabo Chakily, Alzira, Zito Hamela, Crecencia Bento, Anita, Chico, and Dennis, for making my stays in Inhambane pleasant and comfortable. Together, they, and many others that I sadly cannot name by name here, made the project on which this book is based not only possible but also so much fun. I would also like to thank Mozambican friends and colleagues, Euclides Gonçalves in particular, for helping me secure a long-term visa for my family, and Sandra Manuel, for her support in Maputo and in London. I would also like to acknowledge the hospitality of Peter and Elmarie Davidson, who took such good care of us, often for extended periods of time, whenever we were passing through Johannesburg, and to thank Gesine and Judith at Escolinha de Nhapupwe for teaching my daughter Portuguese.

Of the colleagues who have offered guidance during my time at SOAS, I wish to warmly thank Harry West for his careful engagement with my work and unwavering support over the years, as well as Lola Martinez and the late John Peel, who provided precious support and inspiration during those formative years. I am also grateful to Danny Miller, who, as an informal mentor in the early days of this project, introduced me to the "material turn" in anthropology and encouraged me to think through things. I thank Filip De Boeck, Don Slater, and Adeline Masquelier for their careful reading of earlier drafts and for their very useful comments. I also wish to thank Karin Barber, William Beinart, and Heather Horst for their comments on parts of this project at earlier stages. I would also like to record my gratitude to Joshua Bell and Joel Kuipers for inviting me to participate in the "Linguistic and Material Intimacies of Mobile Phones" workshop at the Smithsonian in 2013 and thank the other workshop participants, especially Jennifer Degger, Anna Tsing, and Webb Keane, for reading and commenting on a draft chapter. I also thank David Pratten and Liz Cooper for their insights on uncertainty as part of the "Uncertainty in Africa" workshop they organized at the University of Oxford in 2012. Earlier versions of chapters of this book were presented at anthropology and African studies seminars at several universities, including the London School of Economics, Cambridge University, Oxford University, the University of Edinburgh, the University of St Andrews, Brunel University, University College London, and Harvard University. I thank the conveners and the audiences for their inspiring questions and suggestions. At the University of Chicago Press, I would like to thank Priya Nelson for her enthusiasm and efficiency. I am also grateful to the anonymous reviewers who provided invaluable feedback on how to improve the manuscript.

The African Studies Centre and the Institute of Social and Cultural Anthropology at the University of Oxford have been my academic homes since 2011, and I am grateful to the colleagues and friends I have made in Oxford for their inspiration and encouragement, especially Jonny Steinberg, Sondra Hausner, Elizabeth Ewart, David Gellner, Morgan Clarke, Hélène Neveu Kringelbach, Jocelyn Alexander, Inge Daniels, Andrea Purdekova, Neil Carrier, Elisabeth Hsu, Ramon Sarró, and Marina Temudo. I am profoundly grateful and indebted to David Pratten for his continued inspiration on, and support through, uncertainty.

Of the friends who have helped me along the way, I wish to thank Ak'Ingabe Guyon and Nicholas Green for their attentive reading of earlier drafts. A big thanks to Gabriel Klaeger, my partner in crime and intellectual companion during our years at SOAS. Thanks also to Mike Strack for joining me on part of this journey. I am also particularly indebted to my undergraduate mentor Michel Verdon, whose teaching and friendship have had an enormous impact on my intellectual development. I only wish we could see each other more often. Thanks to my friend and colleague Emmanuel Khan for visiting me during fieldwork and to Pierre Dufloo, my favorite expat in Inhambane, for helping me put things into perspective. Thanks also to Manu Demars for her unconditional friendship, to Rugare Musikavanhu and Felicity Pearce for regularly checking in on me, to Chantal Beaudoin for being so much more than a family friend, and to Alain Biron and John MacKinnon for "taking" me to Africa all those years ago. I would also like to acknowledge the help of Hamo Yedgarian, who kindly looked after Chicken, the cat, whenever I went away on fieldwork and who did everything to make our life in Bloomsbury more comfortable. Thanks also to the guys in the office across the street for keeping me in check all those years.

I also warmly thank my parents, Diane and Georges, as well as my brothers, Louis-Philippe and Marc-Alexandre, and Tante Denise for their love and support. I am extremely fortunate to have such a loving and generous family. Huge thanks for all the summers they spent looking after my daughter in Montreal, allowing me to write and do a bit of fieldwork. This book also marks the start of an exciting new chapter: our return to Canada after a decade in the UK. So thank you for putting up with the distance for so long. A huge thanks also to my godfather, Luc Benoît, for introducing me to Africa via postcards when I was a child. My biggest thank-you goes to my lovely daughter Mia for her companionship, love, and understanding throughout this, at times rather rugged, anthropological journey. It is with love and pride that I dedicate this book to her. I only wish she didn't spend so much time on her phone!

A number of institutions have funded and supported my research. It was thanks to a Leverhulme Trust Early Career Fellowship (ECF-201-443) that I was able to complete the manuscript and work on revisions. Prior to that, a Postdoctoral Fellowship from the Economic and Social Research Council UK (ES/J006831/1) funded follow-up research, relieved me from my teaching duties, and allowed me to concentrate on the manuscript. Field trips to Inhambane in 2012 and 2013 were partially funded by the African Studies Centre and by the Institute of Social and Cultural Anthropology at the University of Oxford. I am grateful for this financial support. The initial field research was funded by the Social Science and Humanities Research Council of Canada (SSHRC) (752-2005-1878), the Overseas Research Students Award, and the University of London (Central Research Fund). I am grateful for the support of these funding bodies. I also wish to thank the Wenner-Gren foundation for awarding me a Hunt Fellowship, which I was sadly unable to accept at the time.

Earlier versions of several passages in this book have been previously published. An earlier version of the prologue appeared as "Being Cool or Being Good: Researching Mobile Phones in Southern Mozambique," *Anthropology Matters* 11(2) (2009): 1–9. Parts of the introduction and of chapter 1 have appeared as "'Travelling while Sitting Down': Mobile Phones, Mobility and the Communication Landscape in Inhambane, Mozambique," *Africa* 82(3) (2012): 392–411. The debate in chapter 4 was originally published as "Breaking up 'Because of the Phone' and the Transformative Powers of Information in Southern Mozambique," *New Media & Society* 13(3) (2011): 444–56. The debate in chapter 5 was originally published as "Mobile Phones and the 'Commercialisation' of Relationships: Expressions of Masculinity in Southern Mozambique," in *Super Girls, Gangstas, Freeters, and Xenomaniacs: Gender and Modernity in Global Youth Cultures*, edited by K. Brison and S. Dewey (Syracuse: Syracuse University Press, 2012), 47–71. I presented some of the key theoretical arguments of this book in an article published as "Cruising through Uncertainty: Cell Phones and the Politics of Display and Disguise in Inhambane, Mozambique," *American Ethnologist* 40(1) (2013): 88–101. Permission to reprint them is gratefully acknowledged.

Prologue

My phone vibrated on the bedside table, notifying me that I had just received a text message. It was 11 p.m. and my husband was already sound asleep. The message was from an unknown number. It read: "I'm waiting for you. I love you." I switched my phone off and lay in the dark trying to imagine who this unknown sender might be. How long would it take for him—I figured the sender had to be a man—to realize that he had sent the message to a wrong number? Would a romantic rendezvous fall through as a result? I wondered. The women I showed the message to the following day all agreed that someone was trying to sabotage my marriage. I was bewildered. Was it not simply a wrong number, as I had first assumed? Most probably. Still, it was not implausible that someone out there was wishing me harm. My romantic image of the anthropologist who is friends with everyone was shattered.[1]

I had only been living in Liberdade, a suburb of Inhambane in southern Mozambique, for a few months when I received that puzzling text message. But so much had happened in this short period of time. First, I had befuddled the neighborhood by not owning a phone. At the time, I saw mobile phones as a nuisance and was convinced that I could live a happy life without one. This was 2006. I had, however, soon realized to my initial dismay that acquiring a phone was going to prove imperative and, for research's sake, I had ended up giving in to peer pressure.[2] To be considered a person by the young people I

1. I first discussed this experience in an article published in *Anthropology Matters* (Archambault 2009).

2. Sunderland (1999) was among the first to promote the phone as a research tool, following her research among jazz musicians in New York. She did so apologetically though and felt she

FIGURE 1. A fresh coat of paint, Inhambane, 2006. Photo by author.

was hoping to understand better, owning a phone was a basic requirement. Although the phone I bought was a simple secondhand Nokia 1110—not quite the flashy handset that my companions had hoped I would opt for—I soon became fascinated with the exchanges that it made possible. Meanwhile, the soundtrack of everyday life in Liberdade was made up of several songs discussing the social impacts of mobile phones, and I also heard these issues debated in everyday conversations. Then, something spectacular happened: almost overnight, the city of Inhambane was subjected to a major makeover. In an aggressive publicity campaign, the fronts of shops and bars, along with decaying concrete walls and a number of decrepit buildings, were painted in either bright yellow and turquoise—the colors of mCel, Mozambique's leading cellular network—or blue, white, and red—the more sober palette of Vodacom, mCel's sole competitor at the time. The visual effect was spellbinding. I read it as an indisputable sign that I ought to take the phone seriously.

Before heading off to Mozambique, I had imagined that my accompanying husband and five-year-old daughter would facilitate my entry into the

had to demonstrate that telecommunication was not necessarily antithetical to the proximity and contact expected of fieldwork. See L. Pelkmans (2009) for a more recent discussion of these issues in an African context.

field. Judging from the experience of other anthropologists who had conducted research in similar settings, I expected their presence to help me build rapport (DeWalt and DeWalt 2002: 62–63). However, things did not quite turn out this way. In fact, I wish I had read *Notes on Love in a Tamil Family* before going to the field (Trawick 1990). Like Margaret Trawick's husband, mine had little interest in following me around while I worked, and much preferred either staying at home—a small house we were renting in the middle of Liberdade—or hanging out with other expatriates.

One of the first persons I met in Inhambane was Benedita,[3] a married woman in her late twenties who had two young children slightly older than my daughter and who lived nearby in a one-bedroom reed house. We paid regular visits to each other and spent numerous afternoons together talking about this and that, often in the company of Isabella, her sister-in-law. More often than not, the conversation turned to intimate relationships. Benedita's husband was an unrelenting womanizer, and Isabella's, a heavy drinker. Both husbands were also extremely jealous and forbade their wives to own a phone, as they feared that this would enable them to contact and be contacted by other men. I listened to Benedita and Isabella fantasize about one day finding better men and detail the different pressure tactics they had developed to try get their way in the meantime. Our friendship was such that they even confided having had secret lovers. This, however, was before my research really picked up, when I was still spending a lot of time at home; when I was still a relatively good wife.

As my networks expanded, the time I spent out and about also increased. Regrettably, this meant that I was breaching core social values in ways almost unfathomable to the people among whom I had elected to live. In Inhambane, a good wife is expected to spend most of her time at home and faces relatively strict control over her comings and goings, in part because the regulation of women's movements is closely linked to efforts at containing female sexuality. Some husbands forbid their wives to attend evening church services or evening classes, even to run a market stall, let alone hang out in male-dominated spaces like *baracas* (small open-air bars). Of course, a wife can sneak out while her husband is himself out or away, but the idea of a woman going out, especially at night, and leaving her husband at home is simply inconceivable. I, however, could often be found at the nearby *baraca*, working into the night—doing participant observation, that is—in plain sight of my neighbors and other passersby, while my husband stayed home to look after our daughter.

3. Some of my research participants asked to have their names changed, but most insisted I use their real names. I was happy to oblige in both cases. In some cases, I have "clouded" some details (Vigh 2006: 20) to prevent easy identification.

As foreigners, we were granted a certain leeway, but still, my behavior, along with my husband's permissiveness—some called it foolishness and lack of vision—prompted great interest and comment, especially from older residents of the neighborhood.

Despite these murmurs, my research was going well; people were opening up to me, involving me in their own personal dramas, and I was filling in notebooks. However, after being attacked one night at machete point—a traumatic incident that nearly cost me my beloved phone—I was forced to rethink my research strategies. That was when I decided to hire a research assistant-cum-bodyguard. After a couple of unsuccessful trials, I eventually met Kenneth, a young man in his early twenties who had recently graduated from secondary school and who lived nearby, not far from the abandoned railway station. Kenneth, who worked as my research assistant for most of 2007, proved incredibly able, reliable, and enthusiastic, and came to play a crucial role in the research. My association with Kenneth not only kept me safe; it also facilitated my access to local youth, as young men from the neighborhood regularly gathered at his house in the afternoons to lift weights and hang out. On the other hand, working with Kenneth provided grist for the gossip mill and rumors about my loose mores soon became ever more contemptuous.

Before long, Benedita and Isabella became my most vocal critics and eventually severed ties with me altogether. Their husbands had allegedly forbidden them to socialize with me as they saw me as a bad influence. The sisters-in-law had, however, proven to me on several occasions their ability to evade their husbands' control—for instance, both had secretly raised enough money to purchase their own mobile phones—and I believe they could have easily maintained our friendship, even as a form of defiance, had they so chosen to. But they were not the only ones to disapprove of me. Whenever I met older acquaintances, it was common for them to greet me politely and then to add "*ah, passear . . .* ," which really meant "oh, so you are out on a stroll [again]!" In contrast, people spoke positively of women, like my neighbor Mira, who could almost always be found at home. The novelty of the *Tsungu* ("European" in Gitonga,[4] the local language) woman living in an all-black neighborhood eventually wore off, and instead of saying "look at the *Tsungu* passing by," residents turned to comments such as "here comes the woman who never stays at home." The very neighbors that were responsible for spreading gossip about me were happy to translate the disparaging comments others made in Gitonga.

4. Gitonga is the main Bantu language spoken in the area.

FIGURE 2. Mia and Omar, learning how to use a mobile phone, Inhambane, 2006. Photo by author.

Although my research was progressing nicely, I was left with mixed feelings. I simply could not shake my romantic understanding of fieldwork.

I started interrogating more seriously, less emotionally, why it was that I provoked such scorn. Little did I know at the time that this reflection would inspire the core argument of this book. I first turned for answers to the classic anthropological literature on gossip. Max Gluckman (1963) understood gossip as contributing to group unity, and, as he later wrote as part of a debate published in the journal *Man NS* (1968: 34), as being essentially about "the evaluation of morals and skills." According to this perspective, gossip acts as a conservative and leveling force designed to keep members of the community in check. By discussing my morally dubious behavior among themselves, it follows, my neighbors were reproducing the system, reminding each other of the difference between right and wrong. But how could Benedita and Isabella be so critical of me when they were secretly cheating on their husbands? What, then, were the "morals and skills" under scrutiny? I could appreciate that spending as much time as I did away from my family raised criticism and suspicion. Having nothing to hide, however, I assumed that by being open and transparent about my activities, I was somehow proving that whatever I was doing, it was all above board. Perhaps gossiping was more about furthering one's individual interests, as Robert Paine (1967) argued in response to Gluckman's analysis.[5] Still, whether gossip was understood as furthering collective or individual goals, the gossip literature failed to provide the closure I was looking for.

My experience of acceptance and rejection gave clarity to the workings of what I will call the politics of pretense. I came to understand that, in this postsocialist, postwar society scarred by profound suspicion, it was not so much *what* I did as *how* I did it that actually mattered and I was, in fact, reminded time and again that "to conceal is respect." It was, in short, my lack of discretion that Liberdade residents found so appalling, as it was seen as a serious lack of respect. As I came to better understand the workings of local regimes of truth, I started using my phone like Mozambicans do, to manage and keep separate different spheres of my life and, ultimately, to maintain appearances, although I never truly managed to reconcile the challenges of conducting ethnographic field research among young people while also being a good wife. The experience also taught me the importance of listening to the field—in my case, "the thing of now" was literally written on the wall, but in other cases the signs may be slightly more subtle—and of remaining open to the possibility, not to say the imperative, of following unexpected paths. This is how I came to write a book about the uptake of mobile phones in Mozambique. What follows, then, is an ethnography of intimacy in a Mozambican suburb through the lens of mobile phone practices.

5. This is a point made by several authors, including Abrahams (1970), Cole (2014), and van Vleet (2003).

INTRODUCTION

Living, Not Merely Surviving

Will the Phone "Save" Africa?[1]

Mobile phones are now part and parcel of everyday life almost everywhere, including in places on the margins of global capitalism. With just over seven billion subscriptions worldwide, global mobile phone adoption is nearing 100 percent.[2] The technology has radically transformed not only how we access and exchange information, but also our very experiences of time and space.[3] It is little wonder, then, that when phones started spreading across the global South, they were invested with transformative potential and sparked great enthusiasm.

In sub-Saharan Africa, where mobile phone penetration rate hovers around 75 percent,[4] the communication landscape has changed drastically since Manuel Castells (2000) wrote, at the turn of the millennium, about "Africa's technological apartheid at the dawn of the information age" (92).[5] Africa is currently the continent with the fastest growing mobile phone adoption rate and is expected to remain in this leading position for the foreseeable future.[6] The spread of information and communication technologies (ICT), and of mobile phones in particular, has even been hailed as "Africa's big success

1. To paraphrase the headlines of an online article by R. Butler (2005).

2. According to the Information and Telecommunication website, http://www.itu.int, accessed February 15, 2016.

3. See Castells 2000; Ling and Pedersen 2005; Ling 2004; Silverstone, Hirsch, and Morley 1992.

4. http://www.itu.int, accessed February 15, 2016.

5. At the time, there were more landlines in Manhattan than in the whole of sub-Saharan Africa (Castells 2000: 92).

6. When I first started researching the adoption of mobile phones in Mozambique a decade ago, I conducted a survey among secondary school students in Inhambane, which revealed that

story—perhaps [its] only one" (Butler 2005: 1), while the telecommunications industry has attracted colossal private and public investment (Thompson 2004: 105) in part driven by the putative link between ICT and socioeconomic development. For the proponents of "ICT for Development," or ICT4D (Donner 2008), mobile phones are vested with the potential to help "save" Africa (Butler 2005). Not only is the phone seen as addressing a "real communication need" (Hahn and Kibora 2008) in places where communication infrastructure is rudimentary,[7] it is also widely believed[8] that mobile phones will help developing countries "leapfrog" stages of development (Muchie and Baskaran 2006: 30–31; Nielinger 2006). The phone, then, stands as both an index of modernization and a driver of socioeconomic development.[9]

I will quickly move on from an analysis of the developmental potential of mobile phones to the more central investigation of how people in Mozambique actually use their phones. However, I first wish to say a few words about the international hype around the enthusiastic uptake of mobile phones throughout sub-Saharan Africa, as it offers a useful backdrop to local commentaries on the dissemination of information, which I explore in detail throughout the book.

There is something about space-time compression that captures the imagination. In a review of the literature on infrastructure, Brian Larkin (2013: 333) writes: "Roads and railways are not just technical objects . . . but also operate on the level of fantasy and desire. They encode the dreams of individuals and societies and are the vehicles whereby those fantasies are transmitted and made emotionally real." The same arguably applies to cellular networks, and, more specifically, to growing access to mobile communication, which also comes with promises of a better world. Space-time compression does, however, also frighten, and new technologies, like the infrastructure on which they rely on to operate, are often "objects of both fascination and terror," as Adeline Masquelier (2002) shows in her research on the expansion of transportation networks in Niger. The developmentalist imaginaries that cast the phone as a panacea to many of Africa's problems are, in contrast, striking for their lack of ambivalence.

---

nearly two-thirds owned a phone. Ten years on, it was nearly impossible to find someone who did not own at least one phone.

7. See de Bruijn, Nyamnjoh, and Brinkman 2009; McIntosh 2010; Porter et al. 2010.

8. I use the word *believe* carefully here to reflect the rhetoric of ICT4D promoters who commonly delve into the register of belief when discussing the transformative potential of mobile phones. Indeed, such potential seems to rest more on wishful thinking than on empirical evidence.

9. See Burrell 2008; Donner 2008; Mazzarella 2010.

In recent years, there has been a progressive shift from the emphasis on the transformative powers of technologies to the transformative powers of information (Robins and Webster 1999: 75), which has percolated through policy and industry discourses. Within this framework, mobile phone communication is bestowed with the potential to enhance socioeconomic development by facilitating the circulation of "useful information," namely business, education,[10] healthcare, and governance-related information (Burrell 2008; Slater and Kwami 2005).[11] For example, Robert Jensen's (2007) much cited[12] case study of a fishing community in Kerala convincingly showed how fishermen equipped with mobile phones gained better access to market prices and could therefore increase their profits. Promoting universal access to the technology in contexts in which other modes of telecommunication are poorly developed, it followed, would boost entrepreneurialism.[13] Other case studies have also shown how widespread phone access could help address some of the shortcomings of an overstretched healthcare system. Innovations in mHealth, a shorthand for the use of mobile communication to support health initiatives in the global South, have included the use of text messaging to remind patients to take their medication (Crentsil 2013)[14] and the use of mobile photography in remote diagnostics (Breslauer et al. 2009).[15] Observers have also celebrated the role that mobile phone communication could play in democratization, namely by helping promote active citizenship, transparency, and accountability. Text messaging in particular has, since the

10. For a report on the potential uses of SMS within the education system, see Traxler and Dearden 2005.

11. See also Melkote and Steeves 2004. For a good review of the ICT for Development literature, see Donner 2008.

12. The article has been cited by 1,033 texts, according to Google scholar, accessed March 14, 2016.

13. The spread of mobile phones has also generated enterprising opportunities directly related to the phone industry, from airtime vendors, phone repair shops, call centers, and a secondhand phone market. At first, the payphone industry grew in parallel to the mobile phone industry on which it depended, but as a growing number of people acquired their own handsets and as competition brought down the costs of mobile communication, what was once a profitable industry is now in decline in many places. Araba Sey's (2011) research on mobile payphone operators in Ghana charts the changing profitability of the industry from its introduction and offers rather bleak prospects for the small entrepreneurs who invested in such business ventures.

14. According to a doctor working at a Johannesburg clinic, the rate of patients abandoning antiretroviral medication has dropped dramatically, going from one in five patients to one in fifty thanks to SMS reminders (CNN 2008).

15. Such efforts have, however, yielded varying degrees of success (Donner and Mechael 2013).

Arab Spring, proven to be a powerful medium of political mobilization across the world (Ekine 2010).[16]

Given the transformative potential of mobile phones, the digital divide is, unlike other divides, seen as particularly problematic and as needing to be urgently bridged through the distribution of affordable handsets, the introduction of competitive pricing, and the expansion of network coverage (Mercer 2004: 51).[17] The consensus is also that the phone should be seen as a necessity, rather than as a luxury item.[18] In reality, however, access in no way guarantees that users will use their phones in development-inducing ways. Yet, efforts to promote universal access tend to overlook this dimension by assuming that once they gain access, users will start accessing (and exchanging) useful information and that development will ensue. Why wouldn't they?[19] R. Jensen's (2007) Kerala study, though based on research in a specific setting, has often been wielded as evidence of the developmental impact of ICT, as though phone practices were independent of context. We may invest mobile phones with transformative potential all we like, but it is in the everyday encounters and exchanges that media practices are given shape and meaning.[20]

The ICT4D agenda is, in many ways, reminiscent of Enlightenment ideals that constructed Africa as, in Stambach and Malekela's (2006) words, "an undifferentiated land where new technology can be used to restructure social relationships and connect people to a global economy" (327). That is to say, it rests on a generic view of poverty in which technological solutions are pre-

16. Aker, Collier, and Vicente (2011) offer a useful analysis of the part mobile communication played in the monitoring of Mozambique's 2009 general elections.

17. As Fillip (2001) put it, "the key concern has been that ICTs have a catalyzing and leveraging potential. The gap in ICTs is therefore more dangerous to developing countries than gaps in access to other things" (quoted in Nielinger 2006: 40).

18. For example, a South African case study unambiguously states that "the role that telecommunications play in livelihoods is critical. Rather than being luxury items that are used to increase one's status and standing, or symbolic capital, mobile phones are in fact crucial to the livelihoods of many poor people in South Africa as communication tools that enable social networks to be maintained, remittances to be transported, and families to remain connected" (Miller et al. 2005: 37). The report refuses to see the possibility of phones being concurrently used for coping strategies and as a source of symbolic capital, as if these were mutually exclusive practices.

19. Medical anthropologists, among others, have long argued that limited adoption of biomedicine often had little to do with restricted access and much more with cultural preoccupations and etiologies (see, for example, Pfeiffer 2004).

20. For more on appropriation see Berker et al. 2006; Buckingham 2008; Hahn and Kibora 2008; Parry and Bloch 1989; and Silverstone, Hirsch, and Morley 1992.

scribed for what are construed as technical problems (cf. Ferguson 1990). The reality is, however, far more complicated. As Brian Larkin (2008) succinctly reminds us, "what media are needs to be interrogated and not presumed" (3). Like older media and technologies, such as the radio, popular theater, and mobile cinema, which were deployed in efforts to educate and modernize Africa (Larkin 2008; West and Fair 1993), the uptake of mobile phones has yielded mixed results.

As I shift my focus toward phone use in context, I am, however, tempted to hold on to the bridge metaphor. When I describe young people as using their phones to cruise through uncertainty, I am referring to a form of bridging, to efforts at addressing the rift between the desired, the expected, and the possible. A bridge deals with a gap, but it does so only on the surface, as gaps usually remain even once bridged over. I will return to this point later when I delve into the politics of pretense in Inhambane.

To the extent that anthropology has engaged with the ICT4D paradigm, it has done so by challenging its main proposition regarding radical transformation with more nuanced ethnographic accounts of phone use in specific settings. Anthropologists are, in fact, particularly good at highlighting continuities in places where other observers are more likely to expect or identify ruptures. For example, some have compared the phone to more traditional technologies of communication such as the talking drum (de Bruijn, Nyamnjoh, and Brinkman 2009), while others have situated the appropriation of mobile phones in relation to local traditions of orality (Hahn and Kibora 2008). Others still insist on situating new media within broader media ecologies. In this respect, Harri Englund (2011: 11), in his lovely monograph on public radio and the negotiation of inequality in Malawi, offers the following word of caution to new media enthusiasts: "In their rush to appreciate the new media, social scientists may confuse what is technologically cutting-edge with what is theoretically innovative. The mundane battery-powered radio can broadcast claims that go to the heart of the intellectual challenges confronting contemporary debates about liberalism and equality."[21]

Heather Horst and Danny Miller (2006) were among the first, in their influential study of mobile phone use among low-income Jamaicans, to challenge the ICT4D framework. They showed how mobile phone communication was contributing to the alleviation of poverty, owing to its role in facilitating redistribution and other coping strategies. Their findings revealed,

21. Williams's (1974) classic study of the social impacts of the television warns of the pitfalls of technological determinism. But see de Sola Pool 1977 for a more forceful assessment of the social impact of the telephone.

however, how poor Jamaicans only occasionally used their phones to engage in income-generating activities. As Horst and Miller put it, Jamaicans relied on mobile phones to *get* money rather than to *make* money. They built on the concept of "expansive realization," originally developed by Danny Miller and Don Slater (2000) in their work on the Internet in Trinidad, to highlight how phones were put to the service of expansive, rather than transformative, objectives, to speak of processes whereby "a new technology allows a previously constituted desire to become realized" (Horst and Miller 2006: 64).[22] More specifically, they showed how Jamaicans were using mobile communication in familiar ways by situating phone practices in relation to the local practice of link-up, which involves the active expansion of social networks. Don Slater and Janet Kwami's (2005) report on mobile phone and Internet use in Accra, Ghana, similarly challenged the ICT4D framework by showing how both were mainly used for everyday mundane communication, rather than to access so-called useful information. Their study revealed that many had never used a search engine—to look up useful information—and were instead using the Internet almost exclusively for social networking.

Many observers would, no doubt, agree with Claire Mercer's (2004) statement that "the recent 'ICT fetishism' of international donors is likely to result in a case of misplaced optimism" (49). Several ethnographically informed studies have, like the ones mentioned above, shown how the link between ICT and development rests more on wishful thinking than on empirical findings.[23] Most would certainly recognize the ambivalent potential of ICT. Indeed, if, for instance, mobile communication can facilitate the monitoring of elections (Aker, Collier, and Vicente 2011), it can also be used to spread ethnic hatred in the run-up to heated elections (Osborn 2008). And while mobile communication offers a convenient platform for political mobilization, governments across the world have shut down text messaging services following periods of civil unrest to curb the use of mobile communication for antigovernment mobilization.[24]

My research in Mozambique similarly calls for a qualified take on the prevailing "techno-enthusiasm" (Robins and Webster 1999: 85), if only because

22. Horst and Miller (2006) were later mildly criticized by some for overemphasizing continuities and therefore for failing to acknowledge actual social transformations provoked by the spread of mobile phones (see Burrell 2012; Tenhunen 2008). See also Mazzarella (2010: 784), who argues that, however misguided, the hype surrounding ICT has nonetheless had "tangible effects 'on the ground.'"

23. See, for example, de Bruijn, Nyamnjoh, and Brinkman 2009; Hahn 2012; Horst and Miller 2006; McIntosh 2010; Molony 2008; Slater and Kwami 2005; D. J. Smith 2006; Tenhunen 2008.

24. See Bertelsen 2014: 9 for a Mozambican example.

individuals often use mobile phones in ways that are far removed from what the ICT4D proponents seem to have in mind. For example, some students reported communicating with colleagues via the phone to coordinate group assignments and to speak to teachers. Most admitted, however, that when contacting a teacher, it was usually within the framework of exchange and reciprocity in which young female students offer, or are coerced to offer, sexual favors for a passing grade.[25] Would the case of young women using their mobile phones to better manage and accumulate "Sugar Daddies" be welcomed as evidence of mobile entrepreneurialism?

Young Mozambicans' eager adoption of the phone has sparked heated debates that speak of competing notions of morality and purpose, especially regarding gender expectations. Although no one would do without it, many offer a frank and sober understanding of the part the phone plays in shady pursuits and, more broadly, in moral and social decay. Several African intellectuals are apprehensive that the spread of the phone will result in the "loss of cultural identity" (van Binsbergen 2004: 111; Thioune 2003). Ali Mazrui and Charles Okigbo (2004) worry, more specifically, that "many Africans are enamored by new communication technologies, very often for *all the wrong reasons*" (16, my emphasis). Unlike in Western contexts where moral panics have mostly revolved around the fear of a loss of intimacy, of digital communication superseding "real" face-to-face exchange,[26] the fear here is precisely the opposite: that phone-mediated communication will encourage a surge in intimacy. There are, however, important parallels to draw with the introduction of landlines into North American homes at the start of the twentieth century. While the phone was welcomed by housebound housewives, it caused trepidation among husbands who worried that the phone may encourage, or at least enable, "illicit wooing" (Fischer 1992). What can be emancipatory for some—the American housewife, in this case—can be particularly threatening for others—the

25. This theme was beautifully developed in the Mozambican film *O Jardim do Outro Homem* (The garden of another man), produced by the Mozambican Sol de Carvalho (2007).

26. One of the worries is that mobile phones, along with other digital technologies, are having a deleterious effect on "real," face-to-face communication (Orlove 2005: 700). Mediation—technological mediation—is welcome, provided it does not alter intimacy too radically. Fears of what we could call the virtualization of sociality have been more or less convincingly downplayed by studies showing mobile-mediated communication was helping invigorate and expand social networks rather than relegating social interactions to the realm of the virtual. Several authors working mainly in European contexts have argued that mobile communication was anything but a substitute for face-to-face interaction. (See, for example, Ling 2008; Vincent 2005). Miller and Horst (2012: 13) have suggested that some of the protest about new media stems from this Protestant desire of unmediated authenticity and subjectivity.

unfortunate husband who fears losing control over whom his wife socializes with. These debates aside, what the Mozambican case makes clear is that people tend to use mobile phones to mute and cover up rather than to expose and contest social contradictions. Crucial to the phone's enthusiastic uptake in Inhambane, I suggest, is precisely the part it plays in helping young people juggle the demands of intimacy, to keep others happy (and quiet) while making their own happiness.

In what follows, while I situate phone practices within sociohistorically specific regimes of truth—within what I call the politics of pretense—I also highlight how mobile phone communication participates, in significant ways, in the redrawing of intimacy underway in postsocialist, postwar Mozambique.[27] More specifically, I argue that mobile communication has opened up virtual and discursive spaces within which new ways of being and relating can be sought and challenged, and within which illicit activities can be pursued with some degree of discretion. Of course, no one is entirely oblivious to what happens under the cover of text messaging, but the discretion granted by mobile phone communication, even in its imperfection, has fundamental implications in a context where people live in close proximity, where privacy is scarce, and where growing socioeconomic inequality is coupled with a widening gap between ideals of respectability and reality, as it helps preserve and reproduce what I call "willful blindness," or modes of not knowing, whether contrived or not, which privilege pretense and concealment over debate and confrontation. In this context, the phone operates as part of a wider arsenal of pretense that includes an assortment of tricks and technologies used to sustain epistemological uncertainty. Mobile communication helps conceal, falsify, and embellish reality, sometimes all at the same time.

It is often through the introduction of new media that new intimacies emerge and that novel ways of thinking and being become imaginable. This is a point that Benedict Anderson (1983) forcefully made in *Imagined Communities*, where he argued that print capitalism had contributed to the emergence of nationalism by encouraging individuals to imagine themselves as part of a wider national community. Habermas (1989) similarly examined the workings of new forms of association through the mediation of communication

27. Mozambique gained independence from Portugal in 1975, after a war of liberation (1962–75) lead under the banner of the *Frente de Libertação de Moçambique* (Frelimo). Peace was, however, short-lived, as in 1977 a nascent guerrilla movement known as the *Resistência Nacional Moçambicana* (Renamo) launched attacks aimed at destabilizing the new regime. These attacks marked the start of a protracted war that eventually mutated into a vicious civil conflict, which lasted until 1992.

technologies (Ginsburg, Abu-Lughod, and Larkin 2002: 5), and inspired a number of authors working on highly diverse subjects, from the global flows studied by Arjun Appadurai (1990) to Karin Barber's (2007) research on the role of textuality in the shaping of new forms of intimacy in Nigeria. Several authors have shown how letter writing, in particular, had fostered the imagination of new modes of privacy and personhood (Ahearn 2001; Barber 2007; Goodman 2009). As I show below, by opening up a virtual intimate space, the phone, like letter writing, and unlike media that essentially inspire through the dissemination of content (Englund 2011)—a process that Brad Weiss (2002) calls "the dissemination of imaginative forms" through mass media (95)—participates in the imagination and performance of new ways of being and relating.

In Mozambique, like elsewhere, mobile communication has transformed how people express love and affection, how and why they argue, and even how they sleep. In this book, I explore the different ways in which young Mozambicans use their phones to craft fulfilling lives and, more specifically, to negotiate the demands of intimacy: the entanglement of obligations, necessities, suspicions, fears, desires, pains, and pleasures that make up intimacy. Much of these efforts depend on a careful juggling of information—not necessarily the kind of useful information that ICT4D promoters have in mind, but important information no less. In fact, better access to information is not always such a good thing. Rather, much rests on one's ability to restrict the circulation of information and, as such, to maintain opacity.

Not Doing Anything in a Mozambican Suburb

Inhambane is a picturesque little town nestled within palm tree groves that stretch as far as the eye can see. Wide avenues lined with acacia trees and decaying art deco buildings stand as reminders of more prosperous days when the city was home to a large population of Portuguese settlers. Today, Inhambane acts as an administrative center and tourist hub. With a population of just over 65,000 inhabitants, it lacks the hustle and bustle, not to mention the anonymity, commonly associated with an urban environment. The city even prides itself for being the cleanest in Mozambique. Yet, for historical reasons, Inhambane is nonetheless a cosmopolitan place. Alongside a sizable Indian community, a growing expatriate population consisting of a mix of NGO workers, South African developers, and other foreigners attracted by the booming tourism industry in the nearby coastal areas, the city has, as a well-serviced provincial capital, also attracted a large number of people from rural areas who fled the hinterland in search of protection at the height of the

FIGURE 3. Liberdade, 2016. Google Earth.

civil war in the late 1980s through early 1990s, or who migrated to the city more recently in search of education or employment opportunities.

About two kilometers from the city center, beyond the abandoned railroad track, where the once-paved road ends and the palm-lined sandy road known as rua Branca begins, lies the suburb of Liberdade. Diverging from rua Branca, one enters a maze of winding narrow paths formed by the tall fences that people like to build around their properties. Most of the houses are made of local materials, such as weaved reeds (*caniço*) and braided palm leaves (*macuti*), though many also have concrete flooring and corrugated iron roofs. A growing number are connected to the grid and have access to running water by way of a free-standing outdoor tap. Most households rely on a combination of petty trade, urban agriculture, and remittances to eke out a living and commonly include members who are working or studying elsewhere and therefore only periodically present. Liberdade is markedly livelier than the rest of the city, even if there is almost a village-like feel to the neighborhood.

I spent much of my time in Liberdade with young men and women in their twenties, though our interactions often also involved younger and older members of their intimate networks. These young adults were born in troubled times marked by a protracted civil war (1977–92) and a failing socialist modernization project. Like their parents who experienced post-independence euphoria (Isaacman 1978), today's youth have high expectations ushered in by consolidated peace and postsocialist neoliberalization (West 2005). Young people today also face a harsh reality, albeit a considerably less brutal one, in which aspirations remain largely unfulfilled. Their frustrations are compounded in a context scarred by an abrupt transition from a wartime economy of extreme scarcity to one characterized by a sudden influx of modern consumer goods (Sumich 2008) and one in which the realization of self has become intimately entwined with consumption and display. Like Mozambicans in other parts of the country (ibid.: 122), the young people I worked with insisted that living conditions were deteriorating, and felt that peace added insult to injury. Samo, a young man who lost his right leg after treading on a land mine when he was a child, provided a poignant assessment of the situation when he said: "The war justified everything, but now that we have peace, how are we supposed to make sense of all this nonsense?" What was nonsensical, apart from growing inequality, was that despite all the postwar reconstruction efforts, Mozambique was still lagging behind with a dysfunctional state apparatus, failing infrastructure, crippled healthcare system, and inadequate educational provisions.

When I first met them in 2006, most of the youths of Liberdade were either studying or had recently graduated from secondary school. The great majority were living under the care of older relatives, often in women-headed households. They were still waiting to reap the benefits of the postwar economy, "to feel globalization" as some of them put it. Although Mozambique was hailed for its economic growth, the labor market had never recovered from major state retrenchment, a reduction of the civil service, and the drop in salaries—hangovers of the adoption of structural adjustment programs in the late 1980s and a protracted civil war (1977–92) (Pfeiffer, Sherr-Gimbel, and Augusto 2007).[28] Despite having followed the career path sanctioned by the government, graduates struggled to find employment. Like other young people across Africa and beyond, they were wrestling to become able and respectable adults, and many described themselves as "not doing anything," as "sitting at home" or "going nowhere." In this postsocialist, postwar economy marked by growing inequality, the so-called crisis of social reproduction (Comaroff and Comaroff 2004) was very much a gendered one. For the young men I had the pleasure to

28. Wages have remained low. In 2009, the minimum monthly wage for a state employee was 1,826 Meticais (MZN), while the basic monthly cost of living per person was estimated at 1,221 MZN (Cooper and Pratton 2015, citing "Salários Mínimos: Propostas Ainda Aquém Das Necessidades Básicas," *O País* [Maputo], 2009, p. 16). In short, it was technically impossible for someone earning minimum wage to support a dependent, let alone a family. In 2007, 50 MZN was worth about $2.25. For comparison, the daily wages of unskilled workers were between 50 and 60 MZN per day.

work with, it was essentially a struggle to live up to mainstream ideals of masculinity as the once attainable ideal of the man as provider was simply out of reach. Young women, for their part, struggled to find suitable partners, and many were second-generation single mothers; but they proved, overall, more willing to redefine themselves. I will discuss, in the coming chapters, the details of these young people's struggles with a focus on the part mobile communication plays in the juggling of everyday uncertainty.

Young people's narratives of stasis deserve to be seen as responses to the specificities of Mozambique's postwar, postsocialist economy,[29] but they also speak of much broader shared experiences. Various authors have recorded similar accounts in other parts of Africa (Langevang and Gough 2009; Mains 2007; Weiss 2005) where educated young adults similarly express and experience their frustrations in spatiotemporal terms (Adams 2009; Hansen 2005). In an attempt to capture the scope and texture of these experiences, scholars have described young people as being "stuck" (Sommers 2012) or in "waithood" (Honwana 2012) in the liminal phase of youth. So pervasive is this crisis that some suggest, echoing earlier prognoses of "lost generation" (Cruise O'Brien 1996; Mbembe 1988), that liminality has become the norm rather than a temporary episode in young people's life courses. However, as Karen Hansen (2005) points out, these narratives deserve to be understood as "discursive metaphors" and should therefore not be mistaken for accurate renditions of young people's everyday existence (9)—that is to say, no one actually sits at home doing nothing. Narratives of stasis speak of expectations measured against a linear understanding of modernization (cf. Ferguson 2002) that prevails both in official discourse and in local imaginaries. In Mozambique, experiences of stasis have emerged as counter-narratives to those of forward and upward movement of the Mozambican nation disseminated through government speeches, advertising, and popular music.[30] These experiences of stasis and desires of mobility are also shaped by Mozambicans' engagement with mobile phones, which are portrayed as both an index and a driver of this forward movement (Rungo 2007).

The crisis of social reproduction is by no means unique to the African continent. Although the shape and form it takes are inflected by local specificities, mass unemployment is commonly cast as the key barometer of the

29. I have explored elsewhere the distinct temporalities emerging out of these young men's protracted exclusion from the labor market and how they commonly delved into idioms of mobility to articulate these experiences (Archambault 2012b).

30. See, for example, MC Roger's hit song "*Moçambique sempre a subir*" (Mozambique always rising) (http://www.youtube.com/watch?v=NdyudQOSMjk, accessed April 22, 2012).

crisis. Telling of the global forces at play in shaping these dynamics, we hear of comparable experiences in locales further afield. For example, Craig Jeffrey (2010), in his discussion of educated jobless youth in India, describes a "culture of masculine waiting." In the introduction to their edited volume, entitled *Young Men in Uncertain Times*, Amit and Dyck (2012) further argue that global economic restructuring has affected a specific category of youth all over the world. This book contributes to these discussions by exploring how young people in Inhambane use mobile phones to navigate everyday uncertainty and the demands of intimacy in their efforts to create fulfilling lives. More specifically, I show how the phone operates as part of an arsenal of pretense designed to cover up some of the contradictions of the postsocialist, postwar economy.

Following the resolution of the armed conflict in 1992, a relatively peaceful postwar reconstruction, and a demonstrated commitment to democratization, Mozambique was hailed as an African success story and became the poster child of the World Bank and the International Monetary Fund.[31] With the turn of the century, interest in the country's reconstruction, which was manifest by the influx of aid, had started waning, and most of the foreign aid was channeled toward mega projects (Castel-Branco 2009). As reports of deepening poverty among large segments of the population started emerging, and as the urban poor took to the streets in protest of rising food prices in 2008 and then again in 2010, observers started talking about a more nuanced Mozambique paradox in recognition of the growing economic disparity underlying economic progress (Bertelsen 2014). More recently, however, the discovery of oil and gas in the north, along with the country's overall resilience during the global financial crisis, have rekindled optimism and attracted a sudden influx in foreign investment. As Mozambique enters an era of prosperity—it now boasts one of the world's fastest growing economies (D. Smith 2012)—it remains unclear how the country's economic growth will translate on the ground.[32]

When they spoke of not doing anything, Liberdade youth were undeniably cynical but nonetheless also resolutely hopeful. "I won't feel [the comforts of]

31. In 2007, the World Bank described Mozambique's economic growth as running at a "blistering pace"; the same World Bank that had, less than two decades earlier, depicted the country as having returned to the Stone Age (Hanlon 2007: 1). Adopting a similar perspective, a report published around the same period suggested that, since the resolution of the conflict, Mozambique had seen a decrease in poverty, coupled with pro-poor development, and that all segments of society had experienced consumption growth (Arndt, James, and Simler 2006).

32. Critical views voiced in the country's independent media, namely the newspaper *@Verdade* (The Truth), reveal a somber portrait of the political culture in Mozambique.

globalization myself, but hopefully my children will," is a comment I heard countless times. Others would simply say: "Hope is the last thing to die!"[33]

Cruising through Uncertainty: The Young Visionaries of Liberdade

When describing how they get by, Mozambicans commonly use the word *desenrascar*, which literally means "to disentangle [oneself] from a situation" (Honwana 2012: xii). Echoing people the world over, they speak of timely improvisation and highlight their resourcefulness in the face of adversity and uncertainty. Several recent ethnographic accounts from across sub-Saharan Africa[34] have highlighted the salience of improvisation in livelihood practices (Vigh 2006), and uncertainty has emerged as a "dominant trope" (Cooper and Pratten 2015: 1). *Uncertainty* refers to a lack of absolute knowledge (Whyte 2009) and is therefore a fundamental dimension of life. Some sociohistorical contexts are, however, arguably more uncertain than others, just like some of us are, for a combination of reasons, better equipped to deal with uncertainty than others. Uncertainty, then, can be understood as an "inevitable force" (Johnson-Hanks 2005: 366), or as a set of parameters that individuals, societies, and species contend with—something of variable intensity that calls for some sort of response even if sometimes the response may actually be no response at all. Uncertainty is therefore not to be confounded with crisis (Roitman 2014). Uncertainty is more subtle, even if sometimes just as momentous; it is also more enduring, even though it may very well be experienced as discontinuous (Archambault 2014). Uncertainty has, in fact, often been conceived of as a form of "structural violence" (Farmer 2004), as the unfortunate product of wider configurations of power that calls for some form of remedial action (Steffen, Jenkins, and Jessen 2005).

There are excellent ethnographies that, drawing on John Dewey's (1929) *The Quest for Certainty*, explore how people engage with uncertainty and the tactics they deploy to regain a sense of authorship over their lives (Vigh 2006; Whyte 1997). Challenging the widespread understanding of uncertainty as a constraining force, scholars such as Elizabeth Cooper and David Pratten (2015) insist that uncertainty should be understood as a social resource, rather than simply as a problem to be solved, or as others have put it, as "a productive force" (Berthomé, Bonhomme, and Delaplace 2012), as something that can be "embraced" (Di Nunzio 2015) and "courted" (Steinberg 2014). In

33. There are interesting parallels to draw between different postsocialist ways of articulating hope (Zigon 2009).

34. And beyond (Amit and Dyck 2012).

the chapters that follow, I examine the ways in which young people in Inhambane use their phones to address everyday uncertainty, to *desenrascar*. I also show, through the lens of mobile phone practices, how uncertainty can produce new ways of being and relating—new intimacies. But this is only part of the story I wish to tell in this book. Were I to stop here, I would be overlooking another dimension of uncertainty that is fundamental to the crafting of fulfilling lives in this part of the world.

Inhambane's particular brand of uncertainty has been shaped by the city's cultural and historical geographies as a settler town scarred by war and socialism, as well as by the region's marginal position in the global economy. It is a material uncertainty—most people do not know exactly how they will be able to make it through the month—but it is also an epistemological uncertainty tied to pervasive suspicion; one that begs endless questions: Is this person trying to cause me harm, and is he or she aware that I am trying to harm them? Can I trust them, and do they trust me? Does this person truly love me, or is he or she only driven by ulterior motives? Is my partner cheating on me, and does he or she know that I am cheating on them? Such uncertainty, in turn, fuels and is fueled by a deep sense that truth is elusive and that all is not what it seems. In this book, I am interested in exploring what François Berthomé, Julien Bonhomme, and Grégory Delaplace (2012) have called "situations of interactional uncertainty"—that is, situations of opacity that generate a sense of uncertainty. I insist, however, on showing how the seemingly straightforward contrast between clarity and opacity is complicated in regimes of truth that, like the ones found in Inhambane, encourage ambiguation over clarification, and dissimulation over revelation and confrontation. In other words, in regimes of truth that privilege open-endedness.

Whether it posits uncertainty as a challenge or as a resource that can be tapped into, as a source of anxiety and hardship, as a source of creativity, or as a force that sets things in motion sometimes for the better and sometimes for the worse, much of this scholarship posits uncertainty as the discrete product of wider configurations of power. But what, then, are we to make of uncertainty that is purposefully generated? What about the uncertainty that is not a by-product but a means to an end, sometimes even an end in itself? In this book, I approach uncertainty not only as a potentially productive force that people engage and contend with in their everyday lives, but also as something that is often deliberately produced. I therefore turn my attention both to responses to uncertainty and to the actual production of uncertainty as a desired outcome. In other words, to uncertainty as *social resource* and as *social practice*. My approach is very much inspired by the works of Marianne Ferme (2001) who, in her study of the aesthetics of concealment in Sierra

Leone, shows through everyday material culture how the production of secrecy participates in the negotiation of power in a profoundly uncertain context marked by a violent past. It is through an investigation of phone practices that I propose to make sense of these entwined aspects. Building on Simmel's (1906) seminal essay on secrecy and secret societies, the anthropology of secrecy[35] offers insight into the workings of uncertainty as social practice. Like the occult knowledge of secret societies that draws boundaries between initiates and noninitiates, the everyday lies that people tell each other similarly participate in maintaining boundaries between the self and the other. This sort of secrecy speaks of a desire to create remoteness (Sarró 2009: 9; see also Davidson 2010: 217). In my efforts to make sense of modes of knowing and not knowing in Liberdade, I build my analysis around the local notion of *visão* (vision).[36]

Although they repeatedly talked about how difficult it was to get by and how they had to *desenrascar*, to improvise and hope for the best, my young interlocutors were, by no means, living in survival mode. As they struggled to get by, they also sought to have a good time, to *curtir a vida* (to enjoy life). "*Curtir* means to live," I was told. Young people's aspirations were, at the experiential level, nicely captured by the distinction they liked to make between living (*viver*) and merely surviving (*sobreviver*).[37] They would have agreed with Nicki Minaj and Drake when they sing: "But to live doesn't mean you're alive . . . Cause everybody dies but not everybody lives!" "Living," "the good life," involved eating tasty food, wearing nice clothes, having mind-altering experiences by watching foreign movies, listening to live music, going to new places, having great conversations, drinking or smoking marijuana, and perhaps having a lover or two on the side, while staying out of trouble. On the other hand, a successful life also meant having something to show for, and, ultimately, being considered a person by others and well as by the state. Specifically, this generally entailed building a house, getting married, having children, and maintaining harmonious relations with neighbors and rela-

35. Essential works include Barnes 1994; Bellman 1984; Ferme 2001; Gilsenan 1976; Gregor 1977; Simmel 1950; Taussig 1999. See also Davidson 2010; Gable 1997; Gonzalez 2010; Murphy 1980; Sarró 2009.

36. An earlier version of the following section appeared in Archambault 2013.

37. With its two degrees of "being" expressed though the verbs *ser* and *estar*, the Portuguese language lends itself to fascinating subtleties, allowing, in some cases, words to go from a positive to a negative connotation. To be *fudido* (fucked) is a good example. *Você esta fudido* means that one is in trouble whereas *você é fudido* means that one is particularly clever. The distinction between *ser feliz* (to be happy as an ontological condition, as a feature of the self) and *estar feliz* (to be happy at a specific moment in time) is arguably subtler.

tives. It involved a careful negotiation of the demands of intimacy. Much of the mobile phone advertising plays on either of these two themes: the phone as a gateway to a lifestyle of leisure, and the phone as a tool designed to keep friends and family connected—two often competing dimensions of the good life.

Liberdade youth described their everyday struggles more succinctly as hinging on one's *visão*, a Portuguese term that literally translates as "vision." Although *visão* actually rests on a much broader sensory awareness, it semantically privileges vision over other senses. *Visão* is, simply put, the poise and cunning required for successful living. It encompasses within it several aspects of urban flair that bring together knowledge, skills, and overall finesse. Like that of the hunters and warriors that Mariane Ferme (2001) writes about, the "survival" of young adults in Liberdade "depends on their own powers of dissimulation, as well as on their ability to identify the traces left behind by prey or enemy and to recognize the guises in which they conceal themselves" (26). Indeed, as others have shown, the skills of the hunter, like those of the pirate (Simone 2006), can also prove invaluable in urban settings (Hansen and Verkaaik 2009: 5; Newell 2012: 80–81).

In Liberdade, where the ground is sand, this is well exemplified by the way in which young people are able to accurately identify the footprints of family members, neighbors, and other acquaintances, and even establish a timeline of their occurrences. In fact, in a sartorial effort at distinction, young people go out of their way to secure distinctive, cool shoes, usually sourced in secondhand markets. These then become quite literally part of their owner's identity, as the marks they leave behind in the sand can be used by others to keep track of their owner's comings and goings. Seen as precious and potentially incriminating repositories of information, footprints are the focus of much concern. Following a robbery, for example, victims collect suspicious footprints found at the scene by scooping up the sand into a container that is then taken to the *curandeiro* (traditional healer) for investigation.[38] A woman I knew could also tell whether her husband had been to the neighborhood where his lover lived by looking at the color of the sand stuck on the soles of his shoes. One learns to tread with caution. On a daily basis, footprints left in the sand by the previous night's activities are read before being swept away at sunrise, when the neighborhood wakes up to the rhythmic swishing sound

38. Footprints can be more than repositories of information to be read. They can also contain part of the essence or lingering presence of the person who left them behind. Among the Walpiri of Central Australia, for example, the footprints of the deceased are smoothed over so as to allow them to become ancestors (Jackson 2013: xv–xvi).

FIGURE 4. Reading footprints in the concrete jungle, Maputo, 2013. Photo by author. Kenneth, my research assistant, who now lives and works in Maputo, still uses his "hunting" skills in the concrete capital city. Upon his arrival to the office one morning, he spotted suspicious footprints, amid other marks, left in the dust in an unused room. Kenneth immediately alerted the security guard who, like me, found it difficult at first to distinguish the strange footprints from the other marks on the floor.

of stiff brooms moving sand around. Reading and sweeping are repeated as people come and go throughout the day and, according to the circumstances, great care is taken not to leave incriminating footprints behind.

Visão is the ability to see and read the landscape, but it is also much more than this. *Visão* enables those who have it to read not only footprints but also the wider social environment while providing the inspiration required to make the most of whatever they encounter. It is an ambivalent quality that can be used for both constructive and destructive ends, to disclose and occlude—two sets of interconnected practices that work in tandem (cf. Taussig 1999) and that I examine in turn. It is also one that individuals possess to varying degrees—in fact, some are seen as lacking it all together—and that is cultivated and enhanced through life experiences. For example, during the colonial period, when the region served as a labor reserve for the South African mining industry, labor migration became a rite of passage into manhood (Sheldon 2002: 3), and those who failed to participate were usually seen as "inexperienced and ignorant provincials" (Harries 1994: 157), in other words, as lacking *visão*. The rationale was part economics, part tied to the acquisition of mind-opening experience and knowledge. In contemporary Inhambane, alternative avenues for developing *visão* include education and involvement with foreigners (expatriates, tourists, and the odd anthropologist), in addition to the consumption of alcohol and marijuana, and the consumption of mass and alternative media. Young people also showcase and develop their *visão* by engaging in a particular form of oral exchange known as *bater papo* (Archambault 2012b; see chapter 1). Individuals recount how their vision was opened (*abri visão*) following a specific event or moment of clarity. Alternatively, one's *visão* can be covered (*tapada*) either by a malevolent force or due to trickery. Along similar lines, describing someone as being blind (*sego*), figuratively, is a great insult that delves into the register of vision to gauge individual social competence.

One with *visão* sees through attempts at concealment. But one with *visão* also knows how to project a certain image, how to play on the visions of others. If those with *visão* get by, by tapping into social networks, by engaging in petty crime, or by exchanging sexual favors, for instance, they also know how to conceal their tracks, by hiding the "ugly things" (*coisas feias*) they do. When Inhambane residents describe something as being ugly, there is often a certain moral assessment in their evaluation. It is, for example, considered ugly for women to smoke, for one to steal or to refuse to give something to someone who asks for it, and particularly to contravene gendered norms, all the more so when men start acting like women and women like men. But there is also a sense that, in a context of growing inequality and profound uncertainty, people are pushed into doing ugly things because of unfortunate life circumstances, and despite themselves. At the end of the day, what is truly ugly is engaging in ugly pursuits without discretion. Finally, *visão* is also deliberately "scotomic" (Taussig 2012: 144), as those with *visão* know when to look away and feign ignorance. Indeed, if being known puts one in a position of vulnerability, so can knowing. And I argue that this is mainly because knowing usually calls for some form of remedial action.

In his research among demobilized soldiers in Guinea Bissau, Henrik Vigh (2006) has put forth a theory of social navigation that sheds light on the negotiation of everyday uncertainty. For Vigh, if navigation is performed in concrete environments—as in the practice of reading and sweeping the sand detailed above—actors also navigate "networks and events" (13). Building on the dynamism inherent in Alfred Gell's model of navigation,[39] while

39. Gell (1985) proposed a dynamic understanding of wayfaring that highlighted the processual nature of navigation. Another important element of Gell's model that others writing after

reinvesting it with a phenomenological sensibility *à la* de Certeau, Vigh nonetheless insists that although social navigation is not exclusively applicable to contexts of social turmoil, it is in such environments that navigation skills are most urgently and visibly deployed. Indeed, the concept of social navigation evokes troubled waters, jerky movements, and a staccato rhythm (cf. Lubkemann and Hoffman 2005); images that do not always gel with the experiences of those living in less volatile environments like the one found in Inhambane, a rather sleepy town, even by Mozambican standards, where everyone knows everyone and where visibility and invisibility, respect and ignorance, are integral to the ways in which individuals make a living.

Borrowing from a seafaring register, young people in Inhambane sometimes talked about being adrift ("*nos aqui em Moçambique estamos à deriva*"), without being moored or steered. Yet, in their attempts to claim authorship over their lives, they deployed much energy to make it look as though everything was running smoothly, as though they were riding the wave rather than fighting the storm or simply staying afloat. Sometimes steering, like in Bernard Shaw's understanding of heaven,[40] often pretending to steer. In short, cruising through uncertainty. And I found that, to embark on this cruise, the phone—a mobile phone—was perhaps the single most important item to pack, to bring along.

The Itinerary

This book draws from field research I conducted between 2006 and 2012 on mobile phone use among young adults in the provincial capital of Inhambane. During the first phase of the research, I had the opportunity to live in the suburb of Liberdade for a period of seventeen consecutive months (2006–7). In addition to semiformal interviews with forty-six young Liberdade residents, I also interviewed figures in Inhambane who were involved, in one way or another, with intimate relationship issues, namely the police chief, the secretaries of various neighborhoods, priests and pastors, and doctors as well as *curandeiros* (traditional healers). I even had the chance to interview the Minister of Youth and Sport, as well as Denny OG, one of Mozambique's

---

him have picked up on is that it considered different forms of navigation—a child walking home from school or a soldier at battle—as distinct in degree rather than in kind. In other words, Gell argued that the same processes were at play whether actors navigated a familiar topography or a more treacherous one.

40. The Irish playwright George Bernard Shaw once wrote: "To be in hell is to drift, . . . to be in heaven is to steer" (quoted in Jackson 1998: 19).

most popular singer-songwriters at the time. Throughout my time in Inhambane, I also regularly asked my young companions to take me through guided tours of their mobile phones. Most of the material presented in this book was, however, gathered through participant observation, through proactive hanging out. This main period of fieldwork was then complemented by shorter follow-up trips in 2008, 2009, and 2012. Between 2012 and 2015, I then spent two months of every year in Inhambane, working with several of my original research participants as part of different research projects. These encounters have continued to shape my understanding of intimacy in this part of the world. Given my ongoing engagement with the field, I often found myself compelled to write in the present tense. Weary of the pitfalls of the ethnographic present, however, I have opted to present much of my material using the past tense. I have nonetheless found doing so, in some ways, an uncomfortable compromise as it appears to freeze the ethnography not in a timeless present, but just as problematically, in a past that once was, at a specific point in time, but that is no more.

If ethnographically grounded studies of new media have encouraged a qualified evaluation of the revolutionary potential of information and communication technologies, this book is more concerned with showing how the study of mobile communication can inform broader anthropological discussions around uncertainty, mediation, intimacy, secrecy, and the making of truth. Both fetishized and the object of suspicion, and as much a tool to negotiate social networks as one used to overcome restraints and limitations, the phone is particularly well suited for anthropological inquiry. Beyond the specifics of mobile phone use in context, this book is also about a broader set of questions that the enthusiastic uptake of mobile phones has provoked—questions about morality, livelihoods, and the crafting of fulfilling and successful lives.

I use this Mozambican example to explore the role of pretense in everyday life, examining the tensions between display and disguise, between knowing and not knowing, that mobile phone practices both index and animate. There are three main ways in which young people in Liberdade use mobile phones in their attempts to live rather than merely survive. First, they use their phones as a marker of membership to the world of those who live life. But the phone is not only a prized possession, it also helps put things in motion, namely by lubricating redistribution networks. Which leads to the second way in which young people use mobile phones in productive ways: to facilitate access to certain people and things. Lastly, they use their phones to play on regimes of truth—to embellish reality by concealing some of the ugly things they do. A look into phone practices reveals how young people experience and mitigate—but also court, produce, and sustain—uncertainty.

The book is an Africanist ethnography that focuses on social life in one suburb in southern Mozambique. But it also speaks to the wider community of anthropologists who have conducted fieldwork on everyday life in other parts of the globe and are shaping debates in the discipline on youth, intimacy, material culture, consumption, personhood, and regimes of truth. Indeed, the story it tells has a wide significance beyond the region. Written in a narrative-driven style that aims to capture the experiences of young Mozambicans, the book zooms in on the everyday lives of a dozen-or-so young adults from the neighborhood of Liberdade. Several chapters also engage with pop music as powerful commentary on the impacts of mobile phones on relationships, gender relations, and intimacy more broadly. Each chapter offers a specific theoretical discussion that is weaved with the ethnography and history of the region. This provides a narrative thread that should allow students, in particular, to reflect in a critically engaged way with the material presented and issues raised, and to get a better grasp of the relationship between ethnography and anthropological theory.

Digital technology fosters imaginaries of the expansion of the spatial and temporal contours of the self and of the social person. How these redefinitions participate in shaping sociality and the circulation of knowledge deserves further attention, especially since, in many cases, this expansion remains a *possibility*. I examine the new forms of dependency and interdependency that have emerged in the postsocialist, postwar economy, and highlight the ambivalent potential of mobile communication as it opens up avenues for advancement while also crystallizing entrenched forms of gender and generational domination. One of the key questions the book addresses is how transformative mobile phone communication is in ordinary people's everyday lives. I ask how mobile communication participates in the imagination and construction of refashioned forms of intimacy, aspiration, and purpose. In short, how has intimacy changed since the introduction of mobile phones? Answering this question is, however, a thorny issue, first because the introduction of mobile phones coincides with profound socioeconomic transformation, and second because the question of change is inevitably a relational and dynamic one that eludes a definitive answer. More often than not, outcomes are ambivalent. For instance, phone use simultaneously builds on socioeconomic hierarchies while also playing a decisive role in the negotiation of socioeconomic disparity, it helps conceal certain things but often reveals other things that are meant to remain hidden, it confers a sense of freedom but can also be a great source of anxiety. Ultimately, mobile communication opens up virtual spaces of intimacy within which new, and not-so-new, ways of being and relating can be tried out and negotiated. I explore how young people use their phones in attempts to circumvent, in gender-specific and contested ways, the existing constraints regarding privacy and intimacy, surveillance and control. Related to these themes is the part the phone plays in the circulation and access to information, especially information that is meant to remain secret.

What follows is an ethnographic investigation of the ways in which young people juggle the demands of intimacy through the lens of phone practices. Ultimately, this is not a book about mobile phone practices per se. Rather, it is a book about young people's attempts to lead fulfilling lives with a focus on the part mobile communication plays in these endeavors. As such, I examine phone practices in their own right *and* as a privileged vantage point into social relations and ongoing socioeconomic transformations, namely into the negotiation of gender relations, generational relations, and the realization of self. In short, as a window into the demands of intimacy. If the phone is what brings the narratives of my young interlocutors together, this book remains an investigation of young Mozambicans' attempts to live fulfilling lives, and if the reader leaves with a sense of what it might mean and feel like to be a young adult in contemporary Africa, then the book will have achieved one of its main objectives. It is through the local notion of *visão* that I propose to shed light on the ways in which young people play with façades—whether phone-mediated or not—in their attempts to craft fulfilling lives. Turned into a useful analytical concept, *visão* links the ethnography to issues of agency and social navigation relevant to the youth scholarship, and allows me to engage with the literature on secrecy through a look at specific ways of knowing, being, and relating.

The Chapters

I start, in chapter 1, by situating mobile phones within Mozambique's wider communication landscape, namely in relation to other communication technologies such as landlines, letters, and the radio, and in relation to a privileged form of oral communication, known as *bater papo*, which offers young people the opportunity to develop and showcase their *visão*. I also explore local responses to some of the challenges of mobile communication in an African context (limited access to electricity, limited network coverage, illiteracy, and financial costs) and introduce some of the solutions designed to mitigate these challenges. I discuss, for example, the practice of sending *bips*, which are intentional missed calls that can mean anything from "call me" to "I'm here." After looking at the hitches and fixes of mobile communication in Mozambique, I conclude the chapter with a critical reflection on appropriation and modernization.

Chapter 2 offers a historical overview of the political economy of display and disguise in the region and shows how the city's particular historical geographies inform the stakes and aesthetics of concealment in a postsocialist, postwar economy marked by growing inequality and material uncertainty. I trace local preoccupations with "being seen" back to the legacies of late colonialism and Portuguese assimilation policies, Frelimo's socialist modernization, and the civil war, and show how young people in contemporary Inhambane juggle visibility and invisibility at a time when they tend to have more to hide than to display. This allows me to situate the phone within an arsenal of pretense that includes tricks and technologies designed to bolster invisibility, such as the cover of darkness, fences, linguistic subterfuges, and collusion.

The crafting of fulfilling lives—the ways in which young Mozambicans cruise through uncertainty—is very much gender-inflected as the options available for young women tend to be rather different from those on which young men can draw, as argued more closely in chapters 3 and 5. While young men can rely on the petty crime economy, young women are, for their part, often involved in the sexual economy. The idea of cruising through uncertainty, here, is obviously a nudge at the sexual component of young Mozambicans' everyday efforts to craft fulfilling lives. These differences aside, however, both men and women have become extremely reliant on mobile communication to tackle and tap into everyday uncertainty.

Many of the handsets used in Liberdade were originally owned by tourists before being injected into the local pool of goods that petty crime further stirs up. In fact, handsets prove so mobile that they can be understood as a form of quasi-currency. In chapter 3, I follow mobile phones as a way into young men's involvement in the petty crime economy at this particular juncture in their lives. The workings of this economy shed light on *visão* as a transferable skill and allow me to contrast states of hypervigilance with two important moments when young people let their guard down: moments of carelessness when they fall victim to crime (*desleixo*) and moments when they blow off steam (*desabafar*). The chapter also includes a reflection on unemployment, consumption, and suspicion in a context where alliances are particularly mercurial, access to wealth highly unpredictable, and trust at best foolish.

These themes are probed further in chapters 4 and 5, which are explorations of intimacy, confrontation, and commodification, with a focus on the ambiguous part mobile phones play in these dynamics. Together, these chapters untangle the articulation between mediation (through mobile communication and through money) and intimacy.

Chapter 4 examines how mobile phone communication has transformed experiences with love, jealousy, and deceit. As mobile communication opens up intimate spaces in which young people can experiment with new ways of being and relating, mobile communication is also understood to fuel intimate conflicts. The chapter shows how mobile communication has transformed courtship practices by making the pursuit of romance more direct and less mediated, as well as by allowing young women to play a more active role, and delves into concerns around authenticity raised by these new affordances. It then turns to the intimate conflicts triggered by the interception of compromising phone calls and text messages, and argues that the nature of these conflicts emphasizes, rather than qualifies, the phone's role as a tool of pretense. The chapter thus offers a mapping of discursive politics in intimate affairs and highlights, more specifically, how these impassioned arguments point to shifting gender and generational fault lines—shifts that are simultaneously enhanced and mitigated by particular phone practices.

The contours of these reconfigurations become even clearer in chapter 5, which turns to the workings of the intimate economy in which sexual services, or the pretense of such services, are exchanged for material gain. Here, phone etiquette acts as a new register to express and address the reconfiguration of gender relations along with subtle and not-so-subtle reworkings of ideas of masculinity and femininity. This chapter starts with a debate on the gender allocation of the costs of communication—in Inhambane, men are expected to cover the cost of communicating with women, who often request callbacks by sending *bips*—and argues that men's worthiness is now gauged by their ability to respond to *bips*. Delving deeper into fears of moral corruption and the ethics of phone practices, these two chapters give voice to a variety of concerns regarding the phone-induced subversion of gendered normative ideals, the commodification of intimacy, generalized suspicion, and a deeply felt crisis of authenticity.

People in Inhambane use mobile phones to foster epistemological uncertainty not only to fool others but also to obfuscate the truth and, in a sense, fool themselves. It is on this last dimension—on what I call "willful blindness"—that I turn my focus in the final chapter. I argue that while the introduction of mobile phones has sparked heated debates, what the ethnography makes clear is that people in Inhambane also tend to use their phones to mute social contradictions and, ultimately, to reproduce epistemologies of ignorance, or modes of not knowing, which privilege pretense over open confrontation. In fact, conflict emerges mainly when the phone fails as a tool of pretense. More specifically, I show how the discretion granted by mobile communication helps preserve an unpleasant "public secret" (Taussig 1999) about the workings of the postwar economy, in which young women are encouraged to exchange sexual favors for material gain at an unprecedented scale. Owing to its discretion, mobile communication helps young women evade being labeled as loose women while also allowing parents and other partners to pretend not to know. Even if the debates discussed in chapters 4 and 5 have opened up discursive terrains upon which to rethink intimacy and gender hierarchies more generally, in chapter 6 we see how the phone occludes more than it challenges, conceals more than it exposes. The chapter engages in a wider discussion of contrived ignorance and the negotiated making of truth, and offers a reflection on the relationship between authenticity—being true to oneself as well as to others—and the rejection of objectification in parallel with concealment, pretense, falsification, and ignorance by looking at how young people use mobile phones to navigate these penumbral topographies. In the conclusion, I reflect on the relationship between anthropology and intimacy and expand on the demands of intimacy as a notion applicable beyond the Inhambane context. I then end with an assessment of the phone's transformative potential.

What I suggest, in the end, is that, in their efforts to craft fulfilling lives and negotiate the demands of intimacy, Mozambicans use mobile phones to maintain open-endedness, to reproduce epistemologies of ignorance rather than to expose and challenge social contradictions—contradictions that are, however, somewhat ironically, enhanced by mobile phone communication. The irony is certainly not lost on Mozambicans who commonly describe the phone as a "necessary evil" (*um mal necessário*). It will become clear in the coming chapters what exactly they mean by this.

1

The Communication Landscape

Billboards, bars, and restaurants painted in the colors of mCel and Vodacom; people texting, talking on the phone, exchanging numbers; phones cradled in brassieres and bulging out of tank tops or dangling from a cord and worn like a necklace: mobile phones have suffused the urban landscape. The new business ventures that have emerged in response to the spread of mobile phones have also transformed the look and feel of the city. Small phone-repair shops selling phone accessories have cropped up everywhere. Young men sporting the colors of either network hover around the main public areas where they sell *recargas*, airtime scratch cards that come in various denominations. High-profile television programs, along with concerts and other major events and national holiday celebrations, are usually sponsored by one of the two companies.[1] One is constantly reminded that a mobile phone is the "must-have" object of the moment (Myerson 2001: 3), that the phone is *the* thing of now.

As I ponder over the statistics compiled by the International Telecommunication Union, I am acutely aware of how quickly outdated they become. The last time I checked, mobile phone penetration rate in Mozambique was nearing 70 percent.[2] It is almost incredible to think that only ten years ago, little more than 3.5 percent of Mozambicans had a mobile phone. And while

1. In Inhambane, there are two such major events hosted on local beaches that now have their place on the calendar of yearly events: the Bara festival, also known as *Verão Amarelho* (Yellow Summer, hosted by mCel) and the New Year's Eve party held in Tofo Beach, also known as *Verão Azul* (Blue Summer, hosted by Vodacom). Since 2011, a third player, Movitel, has entered the scene, though it seems to spend far less on advertising than the other two network providers.

2. Neighboring Zimbabwe, with 80 percent, is slightly ahead, while South Africa leads in sub-Saharan Africa with a penetration rate of 149 percent (according to International

70 percent is arguably a high percentage, in a country where the majority of the population resides in rural areas where limited infrastructure and severe economic constraints inhibit, or at least considerably delay, mobile phone penetration, the figure hides an important "internal digital divide," whereby the great majority of users live either in Maputo or in provincial capitals.[3] According to a survey I conducted in 2007, 71 percent of secondary school graduates in the city of Inhambane were mobile phone owners.[4] During a recent visit to Inhambane, I failed to find anyone in their twenties without a phone.[5]

In this chapter, I situate the entry of mobile phones in Inhambane within the region's broader communication landscape and frame the discussion around some of the practical challenges of using a mobile phone in this part of the world. I show how infrastructural limitations, an innovative industry, widespread poverty, and growing inequality have shaped the ways in which Mozambicans are able to use mobile phones and, just as importantly, the ways in which they choose to use them. The second part of the chapter contrasts mobile communication with other forms of communication, namely with greetings and the popular form of oral exchange referred to as *bater papo*. One of the main aims of this book is to offer an ethnography of phone use in context, and this chapter starts detailing what makes particular phone practices "Mozambican." It will, however, become clear that what also transpires through this ethnography is a common humanity (cf. Jackson 2013: 21).

Hitches and Fixes

The spread of mobile phones has radically transformed how people in Mozambique communicate. In a country with a poorly developed landline infrastructure, mobile phones answer "a real communication need" (Hahn and Kibora 2008: 93). Like any mode of communication, however, mobile communication is not without its challenges and drawbacks. While I will explore in detail some of the moral debates around mobile phone practices in later

---

Telecommunication Union statistics for 2014, www.itu.int/en/ITU-D/Statistics/Pages/stat/default.aspx, accessed February 4, 2016).

3. In 2007, 70 percent of users were living in Maputo and the remainder were mainly concentrated in provincial capitals like Inhambane (interview at the Instituto Nacional das Communicações de Moçambique, Autoridade Reguladora dos Sectores postal e telecommunicações, Maputo, November 13, 2007).

4. The survey was conducted among 320 grade-12 students in 3 different secondary schools in the city of Inhambane. The average age of the respondents was 20 years old.

5. This said, young people occasionally go through periods without a phone, as phone theft is frequent (see chapter 4).

chapters, my focus here is on the materiality of mobile communication and, more specifically, on two of the main challenges that users face, namely how to deal with the costs of mobile communication and how to address infrastructural inadequacies.

It was in 2004 that most of the people I worked with in Inhambane acquired their first mobile phone. Some recalled having gone through considerable trouble to get a phone. For instance, when Osvaldo, who was in his early twenties, moved from Zavala district to Inhambane to study, his schoolmates mocked him because he did not own a phone. Succumbing to peer pressure, Osvaldo managed to get employed in the annual mosquito-extermination scheme, but as a daytime student, he had to cancel his matriculation and apply for the night shift in another school in order to be able to work and study at the same time. Most of the money he earned went toward the purchase of his first mobile phone, while he sent the remainder of his salary to his family back home in Zavala.

If, at first, mobile phones acted as conspicuous symbols of social differentiation, as "weapons of exclusion" (Douglas and Isherwood 1979: 95) that visibly distinguished the "haves" from the "have-nots," with ever increasing ownership, this fault line has waned in importance. Indeed, phone ownership has become a basic requirement for being considered a person, and with the wide availability of affordable handsets, only the most destitute find themselves excluded.[6] The thriving tourism industry in the nearby coastal areas has also had an impact on handset availability. I will return to this particular feature of the political economy in chapter 3.

Many of the phones in circulation are used phones either acquired on the secondhand market, "found" on a night out, or received as a hand-me-down. In fact, "secondhand" is rarely an accurate description of the actual number of hands through which handsets have passed. Some phones have keypads with rubbed out numbers due to overuse, making texting a real trial. Batteries are commonly used well past their normal life span and, as a result, phones tend to require charging more often than newer handsets would, and for those who do not have electricity at home, recharging their phones often calls for resourcefulness. While rural residents may resort to car batteries and solar panels, these solutions are considered inappropriate in an urban context where everyone aspires to have electricity at home in the (near) future, and where acquiring an alternative source of energy would therefore only further distance them from attaining that objective. In the suburbs of Inhambane, the

6. In Jamaica, not owning a phone soon became a sign of destitution, not just poverty. In Horst and Miller's (2006) words, "the phone has become mundane" (64).

number of households connected to the national grid is growing thanks to the implementation of a pay-as-you-go system that renders it more affordable. Those without access to electricity at home usually recharge phones at school or at a neighbor's house, but this involves a number of downsides, namely missed calls, the risk of handset or phone credit theft, and the shame of not having access to electricity at home.

In a phone conversation a few years ago when I was back in London, Kenneth, whom I had been struggling to get hold of, explained that he was going through hard times and that he had spent nearly a week with a flat phone battery since he had no money to top up his electricity meter. Unemployed and out of school, he had no alternative access to electricity. I wondered why he had not asked the neighbor to charge his phone, like he used to before connecting to the national grid earlier that year, but as he explained, "If you don't have electricity at home, it's fine to do that, but once you're connected, it's too embarrassing because it means you're so broke (*tchunado*) that you can't even find 5 MZN for electricity."

Added to these challenges is the unreliability of network coverage. A young man once compared his network to thong underwear: "Now you see it," he said, holding his phone upright to symbolize a woman and then, rotating it 180 degrees to expose her backside, he added, "now you don't!" One afternoon, on my way to Homoine, a small town in the hinterland of Maxixe, I gave a ride to two young men who were standing in the shade of a palm tree on the side of the road. "We are going to Fanhafanha,"[7] one of them said, "to inform them over there about a funeral." He saw me looking at the phone that was dangling from a cord around his neck and, in anticipation, he added, "There is no network coverage over there." "We left from outside Cumbana this morning," the other one said, to emphasize the time wasted on travel. I captured another colorful image at an ancestor worship ceremony I attended in a rural district of Inhambane where there were dozens of mobile phones hanging from the branches of a large cashew tree for optimal network coverage, or "to catch network" (*apagnar rede*). Under what reminded me of a Christmas tree, some cooked while others enjoyed the shade and, in order to attend the beeps and rings that punctuated the drumming, one had to carefully steer clear of the fires, while remaining at the right height not to lose *rede* (network coverage)—precious *rede*!

The phone's main affordance is as much about reaching people as it is about being reachable. In fact, the young people I worked with normally received

7. A locality situated about thirty kilometers from Homoine.

far more calls than they made.[8] The phone provides individuals with some of the advantages of a fixed location, something that can prove life-changing in contexts of migration. Given that mobiles are regularly stolen, however, this fixed location is easily void as few can afford the cost of recuperating an old number. The person living in an area with no network coverage, like the person without a phone, becomes a liability (cf. Plant 2001: 61). For those without a phone, or in between phones, phone sharing offers a temporary solution.[9] Many are, however, reluctant to lend phones for reasons that will become clear in the coming chapters. Indeed, phones are usually jealously guarded as they often contain potentially compromising digital evidence of intimate, sometimes illicit, pursuits.

Prior to the arrival of mobile phones, most Mozambicans had only limited experience with telephones. When I asked how people communicated in the past, I received one of two answers: "We used to walk a lot more" or "We would write letters." As noted earlier, mobile phones were rarely compared to fixed phones as few ever used landlines on a regular basis.[10] Throughout most of the colonial period, letter writing was the only way migrant workers had to remain connected with their families. However, the popularity of this mode of communication was curbed by the high illiteracy rate that prevailed at the time. Unlike South African migrant workers who were actively involved in writing letters by the early twentieth century, it was only in the 1940s that migrants of Mozambican origin started exchanging letters (Breckenridge 2006: 151). Still, the censor working in the 1940s tellingly described letters exchanged between Mozambican migrants and their families as "written by illiterates and very difficult to read" (ibid.). Decades later, illiteracy remained extremely high: in 1960, 94.3 percent of black Mozambicans were illiterate (Cahen 2000).[11]

8. See Skuse and Cousins (2007) for quantitative data on this dimension of telecommunication in South Africa.

9. In response to phone sharing practices, Nokia launched phones with up to seven address books (Corbett 2008: 8).

10. The penetration of landlines stood at the extremely low rate of 0.39 percent in 2007 (interview with Massingue Apala of the Instituto Nacional das Communicações de Moçambique, Autoridade Reguladora dos Sectores Postal e Telecommunicações (INCM), Maputo, November 13, 2007).

11. Another important mode of long-distance communication still used today is radio broadcast. Often used to announce funerals, this mode of communication is only suited to the transmission of public information, not to mention its uncertain outcome. Various television programs are also accompanied by messages usually sent by SMS that scroll by in a reserved area at the bottom of the screen. These messages combine the public dimension of messages transmitted over the radio but, unlike the latter, they tend to convey intimate content.

If illiteracy impeded letter writing, it also limited the uptake of text messaging. All the young people I worked with were literate and knew how to send and receive text messages, but they regularly communicated with individuals who had limited literacy skills as well as only basic knowledge of their phone's functionalities. I often sat with Maria, a middle-aged woman, at her palm wine stall at the market. On what was a particularly slow afternoon, she asked a young man who had a used clothes stall nearby to teach her how to send a text message. Maria had been using a mobile phone for several years but, until that day, she had only ever used it to receive calls and to send *liga-me* messages (generic messages requesting a callback, more on which below). Amazed by how simple it was, Maria started practicing her new skill. The first SMS she sent was to her boyfriend. It read, in badly written Portuguese: "It's me Maria, tell me something nice [*bonito*]." Maria then sent an SMS to her neighbor, "just to say hello." He replied promptly, saying that he was pleased to know that she had finally learned how to send text messages.

Although people commonly compared mobile communication to letter writing, memories of actual written correspondence were often rather fuzzy—when asked to give specific examples of when they had actually exchanged letters, most had to admit that they only faintly remembered having written or received a letter once or twice in their lives—but everyone, young and old, remembered the many "wasted journeys" they had undertaken, and how they often made it to someone's house or place of work only to find that the person in question had gone out. In fact, when rationalizing hefty mobile phone bills, people often pointed out that traveling to another city in order to talk to someone would have been far more time consuming and expensive than making a phone call.

When mCel,[12] Mozambique's state-owned and first mobile phone provider, started operating in the country in 1997, network coverage was limited to Maputo. It was only after 2000, when mCel implemented its prepaid service and expanded its network, that mobile phone use started picking up in and outside the capital. Following the Telecommunications Act of 1999, which set the stage for the deregulation process, Vodacom answered a call for tender and started operating in the country in late 2003.[13] A large number of my

12. The company was first established as Telecommunicações Moveis de Moçambique (TMM).

13. Vodacom is owned by Telkom and Vodafone (UK) and operates in many southern African countries. President Armando Guebuza became a partner of Vodacom through Intelec Holding, of which he is a shareholder. The announcement coincided with the declaration of Vodacom as "100% made in Mozambique" (Carmona 2007: 1–3).

research participants became users at a time when Vodacom had just entered the market and was offering innovative and competitive services. Vodacom conquered many for whom the 50 Meticais (MZN) mCel minimum top-up was prohibitive, by offering a 20 MZN ($1) top-up. Vodacom also appeared at a critical time when handsets were becoming increasingly available and affordable. One could buy a secondhand bottom range phone for 150 MZN ($7), approximately three days of work for an unskilled laborer. From then onward, membership rose drastically.

Today, both operators provide similar services at a similar price. There remains, however, a notable exception: unlike mCel SIM cards, which eventually expire if not regularly topped up, Vodacom SIM cards are good for life. For those who spend variable periods of time without topping up their phones, Vodacom is a more reliable option. More recently, Movitel, a third player, has entered the scene. Like Vodacom, it also offers SIM cards for life along with cheap Internet access.

The pay-as-you-go formula means that even those without a reliable source of income can have access to mobile communication. However, although competition has brought prices down significantly, telecommunication remains relatively expensive. Whenever Liberdade youth would top up their phones, it was usually either for a specific purpose or because they had just received money and wanted to buy credit before spending it on other things. Many even believed that alcohol consumption had decreased since the turn of the century, following the introduction of mobile phones. One young man offered the following illustration: "Before when I'd get hold of 100 MZN,[14] I'd spend it all on beer. Now I put 50 in my phone and I buy beer with the rest!" I must have looked surprise because he added, "Julie, you grew up with a telephone, you had a landline in your house and you grew up playing [*brincando*] with telephones. For us here, it's different; we just started to have the chance to play with phones. And that's why, as soon as we get money, we buy [phone] credit."

As a response to prohibitive costs, users have developed various strategies to keep phone bills low. For example, many opt for per-second billing, which is considerably more advantageous than per-minute billing for short calls. Some will also have their thumb on the hang-up key while calling in order to terminate calls more swiftly.[15] There are also several ways to communicate

14. At the time, 100 MZN was worth about $4.50. For comparison, the daily wages of unskilled workers were between 50 and 60 MZN.

15. Many also take advantage of free calls in the middle of the night. In 2008, both operators started charging 0.029 MZN/minute for these calls. Although still very cheap, this option

free of charge. As detailed below, one of the beauties of mobile phones is that, even without any credit, one can still communicate.

Every day comes with a fresh allowance of ten free *liga-me* (call me) messages, which are messages requesting the receiver to call back the sender. One can also request a return call more informally by sending a *bip*, a practice that involves calling someone and hanging up straight away, before the receiver picks up. *Biping*, or *flashing* as it is known in West Africa (D. J. Smith 2006), has been hailed as a compelling example of creative appropriation. *Bips* are unlimited, so long as the sender has a minimum of credit to actually place a call, but they come with the risk of this credit being "eaten" if the receiver answers before the caller hangs up. Both *liga-me* messages and *bips* can be used for the same purpose—asking to be called back—but a *bip* can also be used as a signal that requires a response other than a phone call.

Often, a *bip* simply means "I've arrived." In some cases, however, a *bip* can take on a more complex meaning and the expected response will usually be preestablished between the sender and the receiver. For instance, for Antonio and some of his classmates, two successive *bips* was meant as a rallying call to meet at the nearby palm wine bar. Or people who live apart may regularly send each other *bips* as a "sign of life" (*sinal de vida*), like a heartbeat, every morning after waking up. In such cases, both parties are able to communicate without incurring any costs. As the following example illustrates, the ability to communicate free of charge is a distinguishing feature of mobile communication.

In the face of the alluring popularity of mobile phone communication, Mozambique's landline provider, *Telecomunicações de Moçambique* (TDM), attempted to reconstruct itself and make the landline experience more closely resemble the mobile phone one by introducing a prepaid service called Blá Blá, which it launched in mid-2005.[16] The package also came with an option allowing for the "personalization of the phone"[17] through the introduction of access codes that turned the landline into an individual, rather than a household, possession, not unlike a mobile phone. More recently still, TDM launched a prepaid service similar to the one just discussed but that could be used with any landline, thus functioning practically like a SIM card. Despite these efforts, TDM could not transpose onto landlines one of the mobile phone's most

is, however, no longer available to those without the means to acquire airtime. On a follow-up research trip in 2009, I found that young people no longer took advantage of this almost free window. "It's annoying to get woken up in the middle of the night," Kenneth explained in a conversation with Jhoker, who added that they had tired of the novelty of free nocturnal calls.

16. In 2006, this system constituted 9 percent of the total lines (Greenberg and Sadowsky 2006).

17. TDM's website, http://www.tdm.mz, accessed April 30, 2008.

important features: the ability to communicate for free. In 2007, TDM ran a television ad which featured a young man playing guitar and who was repeatedly interrupted by his ringing landline. Every time he answered, however, there was no one on the line. The ad was meant to suggest that, since airtime on a landline was so cheap, there was no need to send *bips*; instead, one could simply phone and talk. Everyone agreed, however, that the commercial was seen as a kind of avowal of how useless landlines actually were. Without caller identification, *bips* would remain unanswered.

Communication via text messages also helps keep costs down and most top-ups include a number of free texts. However, as I was regularly reminded, "texting is not two-way"[18] and messages may remain unanswered if the recipient is out of credit. Beyond economics, part of the appeal of text messages lies in their materiality. Texts that are expressive rather than informational are sent and received like gifts that can be kept and reread to activate memories (cf. Lin and Tong 2007). People described how, before going to bed, they enjoyed reading sent and received messages. Many also expressed being heartbroken when having to delete "beautiful messages" (*mensagens bonitas*). We will see in chapter 4 that this was a soft spot that often fuelled intimate conflicts.

The phone analyses I carried out revealed that most of these messages were highly formulaic "Hallmark messages" (Ellwood-Clayton 2006: 360) written by semiprofessional message writers in the business of composing religious, romantic, and/or humorous messages for wider circulation. Many were also gender-neutral and could therefore be passed on by a man to his girlfriend who could then send it to another man, and so on. During the last National Population Survey, one of the popular messages in circulation jokingly accused the recipient of warping the results for being counted in several households. Also popular were text art messages of flowers or hearts drawn with simple characters and that could be displayed on even the most basic handset.

Despite the popularity of mobile phones, most of the communication in Inhambane still took place face-to-face as people carried on visiting each other to inquire about this or that and to properly catch up. The formulaic nature of much of the phone exchanges recalled in form the greetings that are customarily exchanged in everyday encounters, while some of the more creative ones offered a condensed version of a much valued form of oral exchange that demands eloquence and verboseness.

18. These findings contrast with research on SMS in other contexts where reciprocity is expected. Lin and Tong (2007) explore these issues in their research on phone use in Hong Kong.

Bater Papo and the Art of Conversation

A Snapshot of Rua Branca on a Weeknight, Just before Sunset

> Young men fresh out of the shower and dressed in clean clothes gather on rua Branca to catch the breeze and a sight of the girls hurrying home from school. Several young women sitting on beer crates braid each other's hair, some selling roasted corn on the cob. Children play soccer in the street. Here and there, there are little clusters of people looking to *bater papo*.

In Inhambane, much is vested in the oral exchanges that take place in everyday encounters, and a focus on the power of words reveals how vision and speech congeal in this uncertain urban environment. The respect of a hierarchy between generations, as well as between men and women, is taught to children from a very young age. Older individuals are in fact regularly heard complaining about how young people "these days" fail to respect greeting etiquette, though I also often heard these same young people speak of how they conscientiously went about greeting fellow residents as part of a strategy to claim recognition as respectful members of the community.

If mastering greetings is relatively straightforward, given their formulaic dimension, becoming proficient at conversation, on the other hand, is more of an art. Having good conversation is highly valued and being generous with words is arguably as important as being generous with material resources (cf. Abrahams 1983)—so long as the words are good words. *Bater papo* is a Brazilian expression that young people use in reference to a form of oral exchange that could be translated as "to chat," though with the risk of downplaying its social and aesthetic values. As in other forms of oral performances, competence in *papo* sessions is judged on eloquence, knowledge, and creativity (cf. Vail and White 1991: 77). *Papo* can also be used to qualify a relationship. For instance, expressions such as "*bato papo com el (a)*" (I converse with him/her) or "*é amigo-a de papo*" (he/she is a conversation friend) indicate the existence of a relationship that goes further than simple greetings. *Papos* can also acquire a life of their own. For instance, although several years have passed since Papaito's "pool trip" discussed below, young men from Liberdade still mention it from time to time, when reminiscing about the past.

Young people in Inhambane regularly meet with the explicit intention to *bater papo* and commonly engage in meta-conversations in which they assess the quality of the *papos* in which they participate, as well as the conversation skills of particular individuals. They might conclude an encounter by saying something along the lines of: "The *papo* was sweet, we should meet again," "I feel relaxed now," or "Talking like this really relieves stress." During my stay in Inhambane, I organized a number of debates—a format of exchange that students are familiar with—on mobile phone–related issues which proved invaluable if only because of the insight they offered on communication more generally, as they were usually followed by extensive discussions on the debate experience.

Young people also liked to reflect on the productive power that words could have on others. For example, men talked about seducing women with *papo* and used the expression *latar uma dama com papo* (to put a girl into a can with conversation). What is noteworthy about this expression is that it compares the powers of words with those of witchcraft, which is popularized as bending someone into submission by containing their spirit into a container, in this case a bottle (see also Groes-Green 2013). It also links exclusivity with the control over women's mobility and rests on the often-cited distinction between male and female "love organs" whereby men fall in love through their eyes and women through their ears. Some women were, however, just as good at using *papo* to outwit the other sex. Women who took advantage of men under sexual pretenses also relied on *papo* to get them out of uncomfortable situations. As one young woman explained, "If you don't want to give sex, all you need to know is how to talk to close the man's eyes . . . all you need is the right *papo*, all you need is *visão*." There is, however, a fine line between cunning and deceit, and those who fail to deliver risk being accused of having or even being a *papo furado* (perforated/empty talk).

Whereas most people can converse, not everyone has conversation (*ter papo*), and Inhambane residents draw a distinction between *bater papo* and *falar* (to talk). *To talk* is commonly used in the sense of "to gossip" or "to complain." Accusing someone of talking a lot (*você fala muito*) is a great insult. Likewise, *to quarrel*, in Portuguese, is referred to as "to make noise" (*provocar barulho*).[19] *Falar* is sometimes seen as the preserve of women, whereas men are generally portrayed as being better at *papo*.[20] This gender divide builds on wider gendered life experiences. The domestic sphere in which women spend a lot of their time is believed to inhibit the creativity that promotes good

19. Abrahams (1983) shows that in the West Indies being a gossiper was seen as "wrong" but that having nothing at all to say was worse (296). I would say that the same holds true in Inhambane.

20. Timbila performances, which could unite up to fifty xylophones, were a scene of oral expression and competition (Webster 1975; Vail and White 1991) and were also male dominated. The performers were male, and the issues discussed in the songs often revolved around male concerns. Among these, the fear that wives would commit adultery during their husband's absence in the mines was an important theme (Vail and White 1991).

conversation. In contrast, men tend to have broader experiences beyond the household and are therefore seen as encountering more opportunities to develop and cultivate the *visão* on which stimulating *papo* relies.[21]

The preferred *papos* are the ones that bring people to "*viajar sentado*," to travel while sitting down[22]—the *papos* that trigger the imagination and that allow participants to travel mentally and "away from reality" (*fora da nossa realidade*). *Papos*, in short, that conjure up a better world. Throughout my stay in Inhambane, I had the opportunity to partake in a number of *papos* in *baracas*—small bars usually consisting of a stall with an adjoining covered area where customers can sit and which act as important social centers where people meet and pass on the latest gossip—on street corners, at the beach, and at people's homes. For example, on a numbingly hot summer afternoon when no one had the means to get to the beach,[23] I found the usual crowd sitting around in front of Papaito's house. Looking at the recently dug rubbish pit in his yard, Papaito started his swimming pool "trip" (*viagem*) as he detailed how we would build the first public pool in Inhambane. Everyone present took part in the conversation very seriously, contributing to the *papo* by asking Papaito to discuss specific details of the project—"Will there be a cover charge?," "Which materials will we use?"—and emphatically approving his answers. "We'll build the place with local materials, like the South African lodge owners. . . . They build marvels with material that we despise. Epah! How they travel, these guys!" The conversation was done in the present tense, not the subjunctive, and for a good hour we all "travelled while sitting down" in the shade of Papaito's yard.

Like the unemployed Nigerien men that Adeline Masquelier (2013) writes about, who join *fadas*, conversation groups united around tea making and drinking, young men in Liberdade would also spend a lot of their time sitting around talking. In Liberdade, however, the consumption of alcohol and marijuana was seen as a key ingredient of "sweet" *papos*. People place particular emphasis on the sociality of drinking and drinking alone is frowned upon a little like eating alone (cf. West 2005: 37), though for slightly different

21. Women's increased access to secondary education and the "democratisation of drinking" (van der Drift 2002) is helping blur this gender divide.

22. Two famous Chopi composers explained to the musicologist Tracey that one had to "dream" in order to compose music (Vail and White 1991: 123). They were probably referring to an experience similar to the one contemporary young men call *viajar*.

23. The beaches are located fifteen miles from the city and are connected by a regular minibus service. The price of a return trip to Tofo, the main beach, was 25 MTn at the time, the same price as a pint of beer or just under half the daily wages of an unskilled laborer.

FIGURE 5. Jhoker, Inhambane, 2007. Photo by author.

reasons. For one thing, drinks should be shared, but so should what comes out of drinking, that is, sociality and *papo*. Linking inspiration with conviviality, Antonio explained: "You can't drink alone. It'll take you a long time to get drunk, but when you're with someone, your imagination is triggered. Yeah, if you are on your own, you'll be preoccupied [*concentrado*]." In Liberdade, *baracas* were venues of choice to *bater papo*.[24]

Some scholars have approached drinking sessions as arenas that allow and encourage the expression of resistance, as they provide time-out during which tensions are relieved and voiced under the cover of drunkenness (Akyeampong 1996; Cruz e Silva 2001: 37). Also interesting, in my view, is to look at drinking sessions not only as a "break in time" but also as enhancing one's understanding of the workings of society, a perspective put forth by Justin Willis (2002) in his social history of alcohol in East Africa. This hyperawareness ties into a classic anthropological understanding of the reflexivity that emerges from liminality. It was also echoed in young people's narratives and through

24. Young men explained, tongue-in-cheek, that the rising price of firewood was to blame for the move of evening storytelling from the yard to the bar.

their choices of registers when speaking about alcohol consumption. For instance, the expression for having some money to contribute to the purchase of alcohol is *ter idea*, to have an idea. And, while *apagnar rede* (to get network) is an expression usually used in situations when mobile network coverage is limited, it also refers to the level of inebriation that stimulates good conversation. For example, if someone is particularly quiet in a drinking venue, others will say that "he still hasn't got network" (*ainda não apagnou rede*).[25] Drinking sessions almost inevitably include a brief reflection on the social importance of the exchanges they lubricate. As a young man put it, "the passers-by that see us sitting around drinking think that we were just wasting time. They don't realize that we are actually busy talking about important things." *Baracas*, like the Nigerien *fadas*, are potentially subversive spaces from which the contours of reality become more visible.[26]

Mastery of the art of conversation, like phone ownership, is a prerequisite for anyone wishing to claim membership among the "civilized"—a term Inhambane residents are very fond of using—among those with some hope of social mobility. Mobile communication builds on preexistent forms of communication, while also taking on particular qualities tailored to local socioeconomic constraints. A large part of the mobile communication that takes place in Inhambane is, in fact, either coded—thanks to *bips* and "call me" messages—or formulaic and designed to keep phone bills low, often by getting others to subsidize the costs of phone calls. And although some may use their phones to *bater papo*, usually at night when calls are cheaper, most of the *papos* that trigger the imagination are performed face-to-face.

Promises and Enduring Inequalities

The widespread and varied use of *bips* offers a compelling example of creative appropriation. Where one sees resourcefulness and creativity, however, another may see practices like *biping*, climbing up a palm tree in search of a better signal, and relying on car batteries to charge handsets for what they really are: responses to entrenched inequalities in terms of access to resources, opportunities, and basic infrastructure that cut far deeper than the global digital divide. As Larkin (2008) insightfully summarizes, "as the speed of . . .

25. Of course, excessive drinking often has the reverse effect, seriously impeding speech, let alone conversation.

26. Lukacs (2013) makes a similar point in her analysis of mobile phone novels in Japan, which are, however, written mainly by young Japanese women.

life increases, so too does the gap between *actual* and *potential* acceleration, between what technologies *can* do and what they *do* do" (235, emphasis in original). What is truly worrying is that by celebrating creativity, we risk glossing over the deeper politics of necessity. If mobile phones have become essential in claims of membership to what young people in Inhambane refer to as the "globalized world," what this world is expected to look like is inflected by the place's particular sociohistorical geographies. Postwar reconstruction, neoliberalization, increased access to education, and the influx of foreign aid have injected new resources and sparked new expectations that are further shaped by Brazilian telenovelas, Pentecostal sermons, and the tourism industry, to name only the most influential sources of inspiration available for consumption in Inhambane since the turn of the century. From Chinese investors, to South Africans trying to tap into the thriving tourism industry, to young Portuguese revitalizing old colonial economic ties and European oil barons, these new and not-so-new transnational actors have also played a part in reshaping the socioeconomic landscape and, with it, aspirations, expectations, and imaginaries of wealth, prosperity, and progress. The young people I worked with were borrowing from different repertoires to promote new agendas such as gender equality, and, as was often the case, to sustain preexisting hierarchies under slightly different guises. But what this globalized world is expected to entail also rests on a series of shared features, on a broader, more generic understanding of what modernization should actually look and feel like. James Ferguson's (2002) powerful argument against multiple modernities seems particularly relevant in any discussion on technology and infrastructure. As Andrew Ivaska (2011) so nicely shows, along similar lines, in his analysis of cultural politics in post-independence Dar es Salaam, it is precisely modernity's singularity and its "pretense to being unmarked and universal" that accounts for its appeal (28). The phone arguably encapsulates this singularity more than anything else ever has.

Mozambicans commonly describe the phone as *being* development. That is to say that they see the phone as an index of development more than as a driver of development, a view that rests on a widespread understanding of infrastructural development as a defining feature of modernity (Larkin 2013). Whenever a new antenna was erected, the consensus was that network coverage would undoubtedly contribute to the development of the region. By equating the spread of mobile phones with development, they were touching upon the newfound ability to communicate as modern citizens. Yet, alongside the technical challenges discussed in this chapter, they also saw themselves as subverting the phone's potential. People described themselves as beginners, as unable—for lack of experience, cultural proclivities, or a combination of both—to tap into the phone's affordances to the fullest. A middle-aged, educated man put it this way:

> Development comes from these technologies, but they also have many negative aspects. The phone is important because it allows you to know certain things. But then there is a tendency to not want to know things, even if these may be beneficial.

These narratives also featured "the scientists" who had created mobile phones with good intentions in mind (see also Robins and Webster 1999: 68)—intentions and affordances that were understood to be almost inevitably undermined once the technology landed in their hands. I will pick up on this idea of "not wanting to know" in chapter 6 when I explore epistemological uncertainty. What exactly informed this self-deprecating perspective will become clearer as I go into the details of phone practices in Inhambane and examine the part the phone plays in young people's attempts to juggle visibility and invisibility.

2

Display and Disguise

When I asked Inocencio what he made of the hype surrounding mobile phones, he said: "You know, many people who own fancy phones sleep on the floor, but if houses were made of glass, these people would have gotten beds long ago!" As a young man in his early twenties who had recently graduated from secondary school, unemployed and still living under the care of his parents, Inocencio found himself excluded from the world of conspicuous consumption. What Inocencio was suggesting in his moralistic comment was not only a critique of misplaced priorities but also, and I think more importantly, that all is not what it seems in a place where regimes of truth are constructed on a careful juggling of visibility and invisibility. In Inhambane, everyday life involves seeking a balance between displaying enough without revealing too much, between accessing social status and deflecting envy, and between having a good time and preserving respectability, while embellishing reality, often through concealment.

In this chapter, I start uncovering the sociohistorical roots of the political economy of display and disguise within which *visão* is embedded and played out. I go back to the colonial period to show how Inhambane's particular colonial encounter as a settler town shaped local concerns with appearance and pretense. I then turn to the lasting legacy of socialism in today's context of growing inequality before examining some of the incoherencies of the postwar economy alongside a discussion on the relationship between respect, respectability, and concealment. I show how, as people increasingly struggle to live up to mainstream ideals of respectability, the ability to conceal certain pursuits has become all the more important. This, then, is where the phone emerges as a particularly powerful tool in a wider arsenal of pretense.

A Brief History of Display and Disguise among the "Civilized" Bitonga of Inhambane

On his first visit to the area in 1498, Vasco da Gama was greeted by local inhabitants who kindly offered him shelter from the rain.[1] "*Baiete! Baiete! Bela Nhambane, Bela Nhambane*" (Welcome! Welcome! Come inside our houses. Come inside our houses), they said, so the story goes, in Gitonga. Seduced by the warm hospitality, Vasco da Gama baptized the area "Inhambane," "Land of Good People," an attribute that residents hold dear to this day.

These good people are known as the Bitonga, a small ethnolinguistic group whose members share a common language, Gitonga, and number around 170,000 (Gerdes 2001: 115). Neighboring the Bitonga are the Chopi to the south, as well as the Tswa to the north and west, a number of whom have migrated to the city of Inhambane in recent decades. During the colonial period, the Bitonga lived in close, albeit parallel, proximity with a sizeable settler population at a time when the city served as a commercial hub in the littoral society of the Indian Ocean (Alpers 2009). The Portuguese presence isolated the Bitonga from broader political struggles played out in the hinterland (Pélissier 1984: 549–50, 79–89; A. Smith 1973: 580). It also shaped Bitonga language[2] and identity in tangible ways. Bitonga often express a sense of superiority and the conviction that they are the most civilized of Mozambicans, something they attribute, in part, to the privileged relationship their forebears had with the Portuguese.

Following the discovery of diamonds in the Cape colony in 1867, southern Mozambique soon became a labor reserve for the South African mining industry and has been shaped by wage-labor capitalism ever since the late nineteenth century (Cooper 2002: 194; Harries 1994). Labor migration offered miners a modest but nonetheless reliable source of income and became a

1. Despite these earlier contacts, it was only in 1728 that the Portuguese established themselves permanently in the area. In 1761, Inhambane acquired the status of village and settlers initiated the erecting of a garrison. This then consolidated the integration of the region within the slave trade and the broader "littoral society" of the Indian Ocean (Alpers 2009).

2. The most visible outcome of this relative seclusion is a linguistic one. Gitonga grammar and syntax are removed from the main Tswa-Ronga family that spreads across southern Mozambique and includes Shitswa, Shironga, and Shangaan (or Shitsonga). It is, however, syntaxically close (44 percent) to Chopi, a language spoken south of Inhambane (http://www.ethnologue.com, accessed April 13, 2009). In addition to lexical borrowing, various Portuguese grammar rules have also been integrated into Gitonga. For instance, the plural is often formed by adding the suffix *-s* instead of by adding noun class prefixes, as in other Bantu languages spoken in the province.

life-long career choice for a large number of Mozambican men. By the end of the nineteenth century, half of the adult male population in southern Mozambique was estimated to be migrating for work (Harris 1959: 51). These migrants were, however, only semi-proletarianized, as they remained implicated in the rural economy they left behind (Arnfred 2001: 36). Low salaries and the cyclical nature of mining contracts consolidated forms of dependency and interdependency between men and women, and more broadly, between migrant workers and those that remained in the countryside. In the process, working in the mines came to be seen as proof of maleness, even as a rite of passage into manhood (Sheldon 2002: 3). It was in part for this reason that men in southern Mozambique were said to prefer going to work in South Africa even when presented with other work options for equal pay (Felgate 1982: 172).[3] The spread of Christianity and its emphasis on the nuclear family under the authority of a male breadwinner also shaped gender expectations along similar lines, further subjugating women to their husbands (Mate 2002).

Labor migration helped young men emancipate themselves from their elders and provided them with the opportunity to address new forms of obligations, such as paying taxes, and to fulfill new desires such as owning a radio, a bicycle, and a warm South African blanket. As such, it allowed men to consolidate their role as provider—a role that remains a central feature of mainstream ideals of masculinity despite significant reconfigurations of the labor market. To a large extent, wages earned through migrant labor were either channeled toward the formalization of marital unions that further strengthened male authority (Arnfred 2001; Harries 1994), or transformed into assets, if not spent on wine (Penvenne 1995). Rural households therefore continued to depend heavily on agricultural production for their everyday subsistence (First 1983: 9). I will show later how the reliance of many on the income of a few is still a key feature of the contemporary landscape, despite the fact that, following independence in 1975, the terms of the labor agreements between Mozambique and South Africa were altered in response to Mozambique's adoption of international sanctions against the Apartheid regime, drastically reducing the number of migrant laborers (Roesch 1992: 465).

Meanwhile, in urban centers social differentiation was driven by different socioeconomic reconfigurations. In the city of Inhambane, where migrant labor was not seen as a particularly attractive livelihood, there were other opportunities for advancement that proved more appealing to an increasingly cosmopolitan and educated population, namely through various forms

3. It has been argued that access to alcohol partly fueled labor migration (Penvenne 1995: 1–7).

of engagement with Portuguese settlers. Some also found opportunities for advancement within the church (Cruz e Silva 1998). In spite of colonial regulations that granted the Catholic Church a monopoly on education, Protestant missions educated a number of the nationalists who played a leading role in the war of liberation (1962–75) (Cruz e Silva 2001).

Like in rural areas, however, although households also came to depend on money to acquire staples like rice and bread, subsistence in Inhambane continued, in large part, to rest on agriculture, fruit trees, and fishing. Still today, residents depend very much on such sources of food. Most households own coconut trees and grow vegetables for personal consumption, especially different types of green leaves such as cassava and pumpkin leaves that are used to make sauce.[4] Those who have spent time in Maputo or abroad are confronted with the reality that everything when away needs to be purchased, unlike in Inhambane where one can easily gather foodstuff. Both economies bore other important similarities. First, in both cases, many came to expect and rely on the financial contributions of a few wage workers. Second, income acquired through wage work was, to a large extent, used to purchase life's little extras rather than strictly for subsistence. The population came to depend on salaries to live (*viver*), more than to survive (*sobreviver*) (Archambault 2014), a local distinction introduced in the introduction.

From the end of the nineteenth century until 1960 when it was abolished, the *Indigenato* set the parameters used to control the movement and labor of the native population. By forcing all men of working age to be engaged in productive labor for at least six months of the year under penalty of being press-ganged into forced labor (*chibalo*), the *Indigenato* guaranteed a steady supply of labor to the South African mining industry. Then, in 1917, an exception was introduced to the *Indigenato* that would grant assimilation status to a privileged few (O'Laughlin 2000: 13). In order to qualify, candidates had to demonstrate that they were civilized, which entailed mastering the Portuguese language, adopting European ways, namely monogamy and the use of a knife and fork, and engaging in some form of respected work or craft (Marshall 1993: 72).[5] Candidates also needed the support of a Portuguese national, and several ex-*Assimilados* recalled how they had been encouraged to apply for assimilation by their employer while working as domestic servants in Portuguese households. Overall, *Assimilados* benefited from more freedom of

4. The land in and around Inhambane is not particularly fertile, and the region's sandy soil is mainly suited to the culture of cassava and peanuts. Much of the fresh produce is therefore imported either from South Africa or more fertile provinces such as Manica and Sofala.

5. These criteria were legalized in 1917, in Article 2 of Edict 317 (Marshall 1993: 72).

movement than the rest of the African population (ibid.). They also enjoyed some degree of respect, which, in turn, conferred a lasting sense of entitlement. Assimilation status was no doubt difficult to attain,[6] but Inhambane is remembered for its considerable population of *Assimilados*. In fact, several young adults in Liberdade are children of *Assimilados*, albeit, in many cases, the love children of assimilated men and indigenous women who were raised by their mothers alone and have little that distinguishes them from their peers born of "nonassimilated" parents. By the late colonial period, claiming membership to the "civilized world," to paraphrase an ex-*Assimilado* born in the 1950s, came to play an important part in the construction of Bitonga identity as colonial subjects and as a privileged group in the region. However, when Mozambique gained independence, the tables were turned, as those who had collaborated with the Portuguese became the object of suspicion and reprisal (O'Laughlin 2000: 28).

When Frelimo took power, the party implemented a socialist modernization project meant to create a "new man" (*o homem novo*) and, more broadly, to foster a Mozambican identity that would transcend ethnolinguistic differences (Alexander 1997: 2; Geffray 1988: 77–78). This new man was also meant to adhere to a strict moral code of conduct. Frelimo sternly condemned excessive consumption, which it identified as one of the hallmarks of Portuguese colonization (Vail and White 1991: 45–46; see also Penvenne 1995), along with other "vices of the colonial period" including crime, alcohol abuse, and prostitution (Helgesson 1994: 357). The Party promoted instead restraint and sacrifice for the good of the nation (Pitcher and Askew 2006). Frelimo also undertook to eradicate what it called the "internal enemy" (*o inimigo interno*), which included anyone believed to have collaborated with the colonial regime, such as traditional leaders, the clergy, and members of the colonial army; those who followed beliefs and practices deemed reactionary and "obscurantist" such as sorcery, polygamy, *lobolo* (bridewealth), and ancestor worship; as well as anyone who voiced opposition to the Party (O'Laughlin 2000: 28; Vines 1991: 6). The figure of Xiconhoca, a cartoon character who came to life in newspapers, on the radio, and in other party propaganda, came to epitomize the internal enemy (Burr 2010). By personifying inequality and injustice, Xiconhoca was meant to educate by negative example. In the process, Frelimo institutionalized punishment as a state instrument designed to

6. In 1950, only 4,555 Africans of the 6 million living in Mozambique were *Assimilados*, and by 1960, the *Assimilados* represented little more than 1 percent of the African population (Wuyts 1989: 21).

repress, deter, and educate party detractors (Machava 2011). The objective was to make a complete break with the past—the colonial and the customary past.

Frelimo profited from the lessons learned in Tanzania, where independence had come fifteen years earlier, and opted for a model of socialist modernization that promoted the construction of a national identity from a clean slate sanitized of any ethnic specificity. At play in Mozambique at the time, like in post-independence Tanzania, for example, were competing notions of authenticity and modernity under the rule of a party that cast itself as the architect of a moral citizen (cf. Ivaska 2011: 39). Summarized by Samora Machel's catchphrase, "Study, produce and fight," also vividly captured by the Mozambican flag, which features a book, a hoe, a star, and an AK-47, the country's first president wanted to usher Mozambique into a Marxist-Leninist–inspired socialist modernization.

Frelimo fell prey to "Marxist conspiracy theory" (Morier-Genoud 1996: 41) and became highly suspicious, prone to seeing subversion even in the most innocuous acts and obsessing about surveillance and discipline.[7] Alongside Frelimo's appeal to nationalist rhetoric and the engineering of a moral citizen was the party's concern with population control (Alexander 1997: 2). In rural areas, the state attempted, under the guise of service delivery, to round up in communal villages those living in dispersed settlements scattered across the countryside. Just like the fight against the internal enemy, the creation of communal villages was met with varying degrees of hostility and resistance (Lubkemann 2007; West 2005).

Frelimo also adopted a series of policies targeted specifically at youth, a category it saw as a source of both "promise and threat" (cf. Ivaska 2011: 41). Frelimo attempted to tap into youth potentialities but also tightly controlled their activities through various forms of sanctions and state-sponsored surveillance. In an attempt to purge urban centers of unemployed young people, so-called "vagrant youth"—an estimated 30,000 to 50,000—were sent to re-education camps as part of Operation Production, a policy infamously implemented by the country's last president (2005–15), Armando Guebuza, who was Interior Minister at the time (Burr 2010). But the state also gave youth an active role in party politics by mobilizing many through the youth league and by encouraging them to spy on their parents and report any anti-Frelimo

7. Euclides Gonçalves (2013), in his research on local state administration in the province of Inhambane, traces back contemporary practices of secrecy within state institutions—namely public servants' refusal to release public information—to the lasting legacy of Frelimo's fight against the internal enemy and its ethos of suspicion.

activities (Kyed 2008: 407), thus introducing further suspicion within households, some of which were already divided by occult conflicts that had surged as an aftermath of the war (Honwana 2003). At the neighborhood level, *chefes de quarteirão*, or heads of residential units, had a more formal mandate to spy on their neighbors, a privilege they could use for their own advancement or as a form of revenge against those with whom they had grievances (Machava 2011: 608).[8] Mozambicans were constantly reminded that although they had won the war of liberation, the struggle was not over, or, as the party's rallying cry, which was drummed into everyone's mind, had it: "*A luta continua!*" (The struggle continues) (Isaacman 1978).

Frelimo's intentions may have been genuine (Geffray 1988: 81; Isaacman 1978), but they were arguably misguided. When the party converted to neoliberal capitalism under strict structural adjustment constraints, it also retracted its stance on all things traditional and, more recently, officially recognized the importance of traditional authorities in the decentralization of political authority (Obarrio 2010). By marginalizing influential individuals, Frelimo had severely undermined its legitimacy, while feeding mounting resentment that would soon be harnessed by the *Resistência Nacional Moçambicana* (Renamo). Despite its origins as a foreign-backed guerilla movement aimed at destabilizing the new regime, Renamo tapped into growing popular discontent with Frelimo and managed to garner local support, especially in regions that had been historically marginalized (Geffray 1990; Vines 1991; West 2005).[9] The armed conflict between Frelimo and Renamo that erupted shortly after independence eventually mutated into a full-blown protracted civil war (1977–92).

Although the countryside was the scene of violent confrontation, the city of Inhambane remained a Frelimo stronghold throughout the war.[10] In fact, by the early 1990s, a number of urban centers were surrounded by Renamo-controlled areas, thus the image of the conflict as having a leopard skin configuration (Geffray 1990: 220). Many older Inhambane residents I knew remembered the war years, which roughly overlapped with the socialist period,

8. These were dissolved following the turn to multiparty democracy in the 1990s only to resurface by the turn of the century in a nonpartisan form. Today, such groups range from *Vigilança do Povo* (People's Vigilance) to *Políciamento Comunitário* (Community Policing) to *Pessoas de Confiança* (Trusties) (Kyed 2008: 404).

9. The rural/urban divide overlapped an equally important north-south axis. War dynamics in the North of the country where Renamo managed to gain popular support were notably different from those in the South (Geffray 1990; Vines 1991; Wilson 1992).

10. The most brutal clash was the Homoine massacre, which left over four hundred people dead in a single day in the town of Homoine, Inhambane province (Cammack 1987).

more for shortages in consumer goods—epitomized by memories of endless queuing—than for insecurity. As a man born in the 1950s recalled:

> The queuing; it was after independence that it started. When the whites started leaving is when it all started. The Portuguese knew how to get the things to come. . . . The queues, it was the most painful thing we ever had to go through [*Esta coisa de bicha foi a coisa mais dolorosa*]. We were used to going to the shops to buy what we wanted. Then, all of a sudden, we had to go at night to mark a space in the queue [*marcar bicha*] and sometimes not even get the thing we needed. You could spend a whole week without getting bread; you would go to the bakery and not get anything, the ones who were nobodies.

One of the side effects of socialism was that the queuing for rations made consumption highly visible. Undercutting Frelimo's emphasis on equality, it also fed suspicion about social inequalities. As another middle-aged man remembered: "One could queue all day at the bakery and not even get a roll, whereas others never seemed to queue but somehow always had bread." With the intensification of the war in the late 1980s, insecurity in the countryside had significantly disrupted agricultural production and further amplified consumer goods shortages (Alexander 1997). Structural adjustments implemented in the second half of the 1980s then brought a hike in prices, and the country became dependent of foreign aid (Hanlon 1996: 93).

When I first started conducting research in Inhambane, I often heard residents—young and old alike—complain about deepening poverty. Particularly salient in their narratives was nostalgia for "the time of Samora," as the country's socialist period is often referred to in memory of Samora Machel, Mozambique's first president (1975–86); painful memories of queuing notwithstanding. As a measure of frustrated expectations,[11] this longing for a period that many were too young to remember firsthand was even more striking given Frelimo's efforts at what Ann Pitcher (2006) has called the "organized forgetting" of the country's socialist interval (see also Dinerman 2006: 89).[12] Though short-lived, Mozambique's socialist era, which lasted until the late 1980s when reforms were first adopted under the presidency of Joaquim Chissano (1986–2005), nonetheless shaped local imaginaries in lasting ways. This nostalgia spoke of a profound discomfort with the present as some

11. The theme of frustrated expectations has been explored at length in various African contexts (Ferguson 1999; Englund 1996; Piot 2010).

12. Direct references to socialism have been erased from the constitution, the national anthem, and other official documents. Unlike other formerly socialist parties, Frelimo has survived the transition to neoliberalism while retaining power, thus raising the stakes for a smooth ideological transition (Pitcher 2006: 94–98).

even yearned for a mythical colonial past that was, like the socialist period, remembered as one of order, efficiency, and some degree of equality. To substantiate this yearning, one young man proudly told me that the security PIN on his phone was 1498, "the year Vasco da Gama first came to Inhambane!" as he himself put it. I had expected young people to loath all things Portuguese, especially given the brutality of its brand of colonialism. What would Hugh Masekela[13] have made of such a steadfast avowal?

Couched in narratives of decline, what this "nostalgic idealization" (Sumich 2008: 122–23) truly emphasized was a profound discomfort with postsocialist liberalization (cf. Cole 2010),[14] and more specifically, with growing socioeconomic disparity and the new forms of exclusion characteristic of the postsocialist, postwar economy. This was something both young and old recognized even if sometimes through different historical lenses. In the words of Ana, a mother of five born in the late 1950s:

> Life is more difficult nowadays than it was when I was young. Back then there was money but there were no goods. We had nothing to eat, we had to queue but it was more equal. Now living conditions are worse, except for those who have [money]. Now we eat by smell only; by smelling the food that the neighbor is cooking.

Using a stretched out palm to illustrate current social differentiation, she added: "You see, each finger is of a different length, just like there are some who have and some who don't." Ana was speaking for many for whom widening inequality was singled out as "the most painful thing about today's day and age." Growing inequality, as an outcome of political and economic reforms, also shaped local politics of display. On the one hand, those who started reaping the benefits of these transformations had to tread with caution so as to avoid occult reprisal. On the other, as the marginalized came to rely on shady pursuits to get by, they also had to rely on technologies of concealment to keep face. In short, the juggling of visibility and invisibility was driven by an array of concerns.

The stakes of concealment in Inhambane were also further raised by wartime displacement and resettlement. Prior to independence, southern Mozambique was characterized by dispersed residence patterns and the region's populations were said to value the privacy granted by the abundance of land

13. Hugh Masekela, the South African trumpeter and singer, wrote a song entitled "Vasco da Gama" in which he states that the Portuguese explorer "was no friend of [his]."

14. Jennifer Cole (2010) and Henrik Vigh (2006) came across very similar narratives in Madagascar and Guinea Bissau, respectively.

(Felgate 1982: 28; Junod [1912] 1966: 296–97).[15] Older Liberdade residents recalled a time when land in the neighborhood was so abundant that "you couldn't even see your neighbor." However, at the height of the civil war in the late 1980s, many people living in rural areas fled to urban centers in search of protection. At first rural residents found refuge in communal villages or district capitals, but as the conflict escalated, many were forced to move, often for the second or third time, to bigger cities like Maxixe and Inhambane. Suburbs such as Liberdade underwent a complete spatial restructuring to accommodate this influx.

Jhoker Macuacua's experience of resettlement illustrates this well. Born in 1982 in a rural area about twenty kilometers from Inhambane city in the neighboring district of Jangamo, Jhoker and his family had relied on agriculture as well as on the income of Jhoker's father, who worked as a miner in South Africa for most of his adult life. When the war reached the area in 1989, the family sought refuge in a nearby communal village, but as Renamo troops continued their advance, they had to resettle closer to the city and further from their fields and coconut trees on which they continued to rely for subsistence. When Renamo soldiers attacked the cashew factory near their new residence, the Macuacuas decided to move yet again, this time to Liberdade, not far from where Jhoker and his family were residing when I met them. Following the resolution of the conflict, some returned to their rural home areas, but many, like the Macuacuas, opted to remain in the city as their houses had been destroyed, animals killed, and fields overgrown, and they had formed new urban networks (Chingono 1996; Macamo Raimundo 2005). Jhoker's younger brother, Augusto, even spoke positively of the war, which, in his understanding, had freed him and his siblings from the "backward isolation" of a rural lifestyle. "If it wasn't for the war," explained Augusto, "I would probably still be living in the bush, I would already have I don't know how many children and I would be surviving off subsistence farming. Instead I'm about to graduate [from secondary school]!" And so, by the early 1990s, Inhambane's suburbs had become densely populated with displaced individuals, many of whom were, like Jhoker's family, of Matswa origin.

15. Writing at the beginning of the twentieth century, the missionary and ethnographer Henri Junod ([1912] 1966) wrote: "Thongas like to build among trees, to protect themselves against the terrible south winds which frequently blow across the plain, and perhaps also to shield themselves from the inquisitive eyes of people passing along the road. The little community [merely an enlarged family] prefers to live by itself, as it is entirely self-sufficient." See also Evans-Pritchard (1976: 13) for a similar observation in Zandeland, but with more explicit references to the fear of witchcraft and adultery associated with proximity.

In this reshuffling of populations and concomitant reorganization of urban space, coconut trees emerged as important stakes in the region's political and economic struggles. To start, the local diet is built on coconut milk, which provides high nutrition and calorie content, especially when combined, as it often is, with groundnuts. Coconut milk is used to cook anything from several kinds of greens that grow in abundance and that are particularly suited to urban agriculture, to seafood and fish or even chicken and meat. But coconut trees are so much more than a source of food: tapped, they produce palm wine; chopped, they turn into planks; and pruned, the palms are used like thatch to build fences, walls, and roofs. The Macuacuas' land was close enough to Inhambane for them to go back every now and then and collect coconuts for consumption in the city. Other families had to travel far greater distances and often had little choice but to purchase coconuts, unlike Bitonga families who could rely on their own nearby trees.

As a valuable resource, trees are also directly tied to land tenure (O'Laughlin 1995: 101). Since land was nationalized shortly after independence, only the right to usufruct the land can be bought and sold, as the land itself belongs to the state. Trees, in contrast, are individually owned and de facto determine land ownership across the region. If the Bitonga owned most of the coconut trees and therefore controlled trees and land access, the displaced were, for their part, more likely to have relatives involved in migrant labor on which they could rely for remittances. Some also worked as domestic servants while others turned to petty trade. Several Matswa families did, in fact, turn this coconut tree setback to their advantage. As others have argued, wartime insecurity in urban settings "unleashed latent entrepreneurialism" by forcing individuals to develop survival tactics and, in turn, contributed to the emancipation of young people from the control of male elders (Chingono 1996). Although the Matswa were initially despised, to varying degrees, by the urban Bitonga for being "from the bush," many now occupy privileged positions in the local economy.

In this period of resettlement, language became a key marker of difference. Zuba, a Bitonga born in Inhambane in the 1960s, explained, "And we were told that the Matswa were inferior, that they were people from the bush who didn't know how to speak Portuguese." He recalled how his parents forbade him to speak Gitonga and forced him to speak Portuguese when he was a child.[16] Today, Inhambane's population reflects the ethnolinguistic diversity of the

16. A number of children are growing up without knowing how to speak Gitonga, as parents prefer speaking to their children in Portuguese, in part to give them a head start with primary education, which, until recently, was taught exclusively in Portuguese.

province (Cahen 2004: 94), and Bitonga, Matswa, and Machopi[17] cohabit in relative harmony. Although they rarely intermarry—Taninha, a young woman of Bitonga and Ronga[18] ancestry swore she would never even date a Matswa—they live in the same neighborhoods, attend the same schools, pray in the same churches, drink in the same bars, and engage in *bater papo* together. The young adults I worked with were fluent in at least one African language (Gitonga, Chitswa, or Chichopi), which they commonly privileged to communicate with kin, and used Portuguese as a lingua franca.

The city's religious landscape is almost equally divided between Christian denominations and Islam and also characterized by peaceful cohabitation.[19] There is also a substantial Hindu community comprising mostly merchant families that originally came from Diu, India. The Catholic Church has a strong historical presence in the city, and the majority of my research participants defined themselves as Catholic, though most only rarely attended church, if ever. Inhambane had not escaped the proliferation of Pentecostal churches, but their presence was marginal when compared with other provincial capitals,[20] something I would attribute in part to the importance of the Catholic Church, its relatively light demands on members (Macamo 2005: 89), and its standing as European import. The Muslims I worked with, for their part, had an essentially orthopractic approach to religion whereby their identification to Islam was commonly expressed through the respect of certain key prescriptions, fasting, and the celebration of Eid. Overall, religion was not a particularly salient feature of the everyday lives of the young adults I followed.

Bitonga, and sometimes Inhambane residents by extension, are referred to by other Mozambicans as Mainhambane,[21] a derogatory appellation which hints at an unspecified Bitonga essence. For example, people would say, "Oh well, Mainhambane . . ." to explain just about anything, from an unusual murder to a shop going bankrupt, without completing the sentence, as if there were a shared understanding of what it entailed to be Mainhambane. Some

17. Individuals of Chopi origin, on the other hand, were usually described in less condescending terms. The Machopi are renowned to excel in several areas valued by the Bitonga, namely they are recognized as possessing special intellectual capabilities that make them succeed in school.

18. Maronga are an ethnolinguistic group from the Maputo region.

19. According to the 2010 population census (http://www.ine.gov.mz, accessed December 1, 2010).

20. In contrast, Pentecostal churches in Chimoio, the capital of Manica province, have proliferated dramatically since the 1990s (Pfeiffer 2006).

21. In a literal sense, the term is the plural of Inhambane, as "Ma" is the Bantu prefix for plural nouns in the human noun class.

Bitonga were themselves using the expression in their introspections on Bitonga cultural heritage and, more specifically, when voicing concerns with "seeing and being seen" (*ver e ser visto*). In some cases, they attributed their preoccupations with appearances to Christian morality and its gendered tenets on respectability (Sheldon 1998). More explicit, however, were ideas grounded in a colonial legacy whereby the Portuguese were remembered as lesser colonizers driven by hedonism rather than imperialism (see also Hanlon 1996: 10). A young man put it to me this way:

> The Portuguese and the British were very different colonizers. Unlike the British whose main concern was to invest, the Portuguese were more preoccupied with looking good and having a good time . . . and that's where our materialism comes from!

In fact, many Bitonga readily recognized that their claim to a superior status, inherited from rubbing shoulders with settlers during the colonial period, had an unpleasant side, one that could easily be taken for snobbery and which rested on a potentially explosive mix of condescension, spectacle, and artifice.

Being Seen in Contemporary Inhambane

In Liberdade, where land has become scarce and expensive, residents now live in close proximity. Every morning, blaring sound systems take over—the louder, the better—as neighbors compete in decibels. Music sets the tempo of everyday life in the suburbs and also plays an important role in muffling conversations in this context of aural proximity in which neighbors and passers-by not only smell—remember Ana who complained about eating by smell only—but also see and listen to each other, as most domestic activities like cooking, washing clothes, eating, and conversing are performed outdoors in the yard. Some days, this social intimacy with its "objectifying gaze" (Sartre 1943) translates into a failure to impress due to excess familiarity. Young men often complained that because Inhambane was so small, people were "tired of seeing each other." As residents liked to say, "In Inhambane, there is no exit" (*não há saída*), an expression of capitulation that referred, in part, to the city's lack of opportunities, in part, to the cul-de-sac location of the town, which was built on a peninsula.[22] What they meant by this was that since everybody

22. There are three ways into the city: taking the thirty-kilometer turnoff from the National Highway, hopping on a dawdling overcrowded ferry across the bay from Maxixe—talks about building a bridge that would connect the two cities have recently been revived; however, the

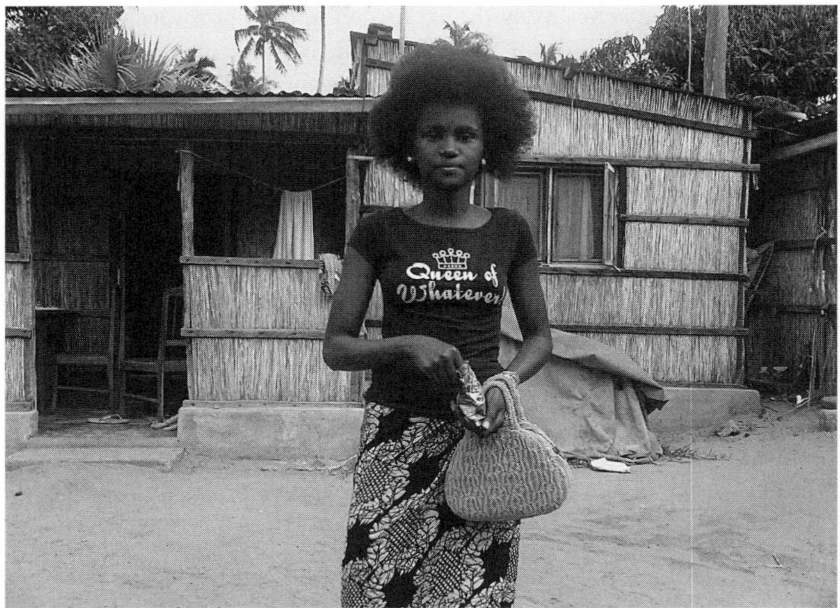

FIGURE 6. Neidi, Inhambane, 2012. Photo by author.

had seen everyone else so many times already, even those with style struggled to stand out and turn heads. On other occasions, however, this social intimacy felt like living under constant and potentially harmful scrutiny. If surveillance in the 1980s served a specific political agenda, surveillance in the postwar context, or the fixation on "the neighbor's life," as it was colloquially referred to, remained tied to suspicion, distrust, and attempts at playing on inequalities.

Like their parents, who were contenders for assimilation, Inhambane youth aspired to gain higher status. This time around, however, the prerequisites were more vaguely defined and acceptance not as definite. The political implications were also quite distinct. Membership was crucially acquired through distinction: young adults in Liberdade liked to emphasize their individuality, their uniqueness, through oratory skills (as discussed in the previous chapter) as well as through sartorial idiosyncrasy.[23] Unlike the Congo-

---

presence of shifting sands in the bay's waters makes the construction of a bridge structurally difficult and, according to experts, highly unlikely—or catching a flight that connects the city to Maputo and Johannesburg.

23. Local concerns with display were also manifest in the ways in which domestic space was kept. Built out of local materials, many of the houses in Liberdade were falling apart for lack

lese sapeurs (Friedman 2004) or *nouchi* youth in Abidjan (Newell 2012) whose sartorial efforts at distinction rest on the combination of branded clothing and accessories, young people in Liberdade relied mainly on special finds in the city's secondhand clothes markets. For a time, Papaito, who liked to clad himself with all white garments, an intrepid option in such a dusty place, even dated a slightly older woman who was selling secondhand clothes at the market. As she gave him and his friends first pick whenever she received new merchandise,[24] the young men in the neighborhood forgave him for dating a woman who they described as "*cansada*" (literally "tired," i.e., no longer fresh).[25] A few winters ago, Jhoker sported a fur-lined checkered hunting hat with ear flaps that could be worn tied up or down, and Kenneth had a hand-knitted jumper in pastel shades with cable detail that would have looked ridiculous on anyone else. Tanhia had a low-cut yellow top that she liked to wear with a pair of black shorts and, to complete the look, she would smear her eyelids with the bright red lipstick I once gave her as a birthday present. Jenny often wore a little gingham dress reminiscent of the uniforms worn by maids in South Africa and that she would button all the way up in a way that seemed to clash with her dread locks and overall laidback demeanor, and Lina liked to cut out slits in her t-shirts and trousers, and then wear these over tight-fitting garments with contrasting colors. Like other young people all over the world, from unemployed men in Arusha (Weiss 2005) to middle-class South Asians (Sharma 2010) who identify with Black American hip-hop culture, young people in Liberdade were also creating their own unique styles in ways that spoke of similar claims of membership.

When phones first started trickling into Inhambane, they acted as tangible proof of membership to the civilized world. Like the jeans, bikes, and sneakers young Indian men wear to "shine" (Lukose 2012), phones were seen as the veneer of civilization that conspicuously distinguished "those who live" (*os que vivem*) from "those who merely survive" (*os que apenas sobrevivem*).

---

of resources to rehabilitate them. Yards were, however, kept impeccably clean, carefully swept at sunrise every day, as mentioned earlier, and many were nicely landscaped with ornamental plants (Archambault 2016).

24. Secondhand clothes sellers purchase their merchandize in tightly packed bales of clothing usually sorted by type (Brooks 2012: 191).

25. Clothes are individually owned and should not be shared mostly because curses can be transmitted through clothing. Nowadays and to the great dismay of older generations, clothing is sometimes shared among siblings and even friends. Fashion shows are very popular in Inhambane. Each model teams up with a *madrinha* (godmother) who is responsible for selecting the model's attire. A good *madrinha* is one who is well connected and able to borrow clothes and shoes from other girls.

As Inocencio put it, "You see, we collect phone numbers in case [we need to contact these people] but also to ascertain that they actually own a phone." It could indeed be argued that young people have harnessed the phone for self-fashioning, as the phone is used to express aspects of one's identity, especially in terms of status. The young people I worked with were extremely knowledgeable of phone model specifications and prices. Most could, however, only afford bottom or middle of the range secondhand phones that, as mentioned in the previous chapter, were often faulty and far from flashy. I often heard young people ask rhetorically, "Why do we need fancy phones if the objective is communication?" as they assessed the economics of mobile phones through a folk model of rational choice theory in which prioritization was sifted through a moralistic sieve. Yet many of these same young people would eventually admit that, were they ever to come into money, they would definitely invest in a superior model.

More often than not, because they were of lesser quality, handsets revealed information that the owner would have preferred to keep hidden, such as the fact that they were too poor to buy a phone that worked properly and that was not falling apart, let alone one that looked flashy. If handsets were not always good enough to be used in the performance of redefined identities, phone practices seemed more powerful vectors of self-fashioning. Janet McIntosh (2010) has also picked up on this affordance in her analysis of text messaging in Tanzania, in which she suggests that linguistic choices young people made when texting reflected their efforts to reconcile the respect of Giriama tradition with the profound desire to partake in globalization.

Much of what I am describing here ties into broader discussions about performativity, compromise, and the display of potential. Within the recent scholarship interested in the performativity of everyday life (Pratten 2008a; Prince 2006; Vigh 2006; Weiss 2002), I find Sasha Newell's (2012) work on the bluff in Côte d'Ivoire particularly inspiring. *The Modernity Bluff* explores the lives of young, mostly unemployed *nouchi* youth in Abidjan who, despite struggling to get by, engage in the bluff, that is, lavish, potlatch-like performances staged in the city's open air bars or *maquis* where, dressed in the latest designer clothes, they spend large sums of money on food and drink. Newell argues that such conspicuous consumption deserves to be understood as "a display of potential" (1), as "a positive transcendence of [one's] surroundings" (5), rather than as acts of deception. In other words, the bluff is, Newell insists, explicitly a bluff; and like a good bluff, it has tangible effects. He writes: "The unarticulated knowledge that the bluff does not index real wealth, allowed the performative act through which *nouchi* transformed the symbolic capital of their street name into the social capital of fistons, producing real success behind the trick mirror of its imitation. The machinations of the bluff provided a symbolic solution to the paradox of reputation—at once demonstrating one's proficiency at deception while at the same time distributing wealth and goods to the significant members of one's network to demonstrate one's worthiness" (98). In a Liberdade *baraca*, like in an Ivoirian *maquis*, performance is powerful when it convincingly conjures, as Tsing (2000) puts it, "a world that is sweeter" (119). Michael Taussig's study of plastic surgery in Columbia—or what he calls cosmic surgery—speaks for a similar economy of appearances; a world in which the display of large fake breasts will hopefully attract the attention and favors of powerful drug lords. Performance, simply put, is potentially productive, even when those involved are themselves aware that it is precisely that: a performance. Through this conjuring, no one is duped. There is no denying that the breasts are fake, that youth in Abidjan are bluffing, that Inhambane youths' cruise is in fact strained. Uncovering the truth, the actual truth, would, however, be beside the point.

New media have opened up virtual spaces of sociality within which such potential can be expressed and tested. Caroline Humphrey (1999) has shown, for example, how participants in Russian chat rooms were revealing through their avatars elements of their personality that would otherwise remain hidden in everyday life. She notes: "The whole ideology of the avatar is that it reveals to the world the aspects of the self that are suppressed in ordinary life because of conventions or the gender/age/status pressures of family relations" (46). More recently, Boellstorff (2008) has similarly suggested in his ethnography of the virtual world *Second Life* that dwellers in this world felt that they could address certain challenges or contradictions they faced in the actual world and "become closer to what they understood to be their true selfhood, unencumbered by social constraints or the particularities of physical embodiment" (121).

It is precisely a latent potential that I have in mind when I evoke the idea of cruising through uncertainty. By juggling visibility and invisibility, my young companions were showing to the world who they could be, who they wished they could be, who they may one day become if the stars ever aligned. In order to render visible their latent potential, they carefully manipulated regimes of truth, simultaneously relying on simulation and dissimulation. I gave several examples of local phone practices in the previous chapter and detailed the social competence required. Overall, however, I found that young people in Inhambane were using mobile phones more for disguise than for display. Although, phone model permitting, ringtones were carefully chosen and updated so as to impress, many nonetheless preferred to keep their phone switched to silent mode in order to evade the scrutiny of others who might inquire about incoming calls and text messages, a point of contention examined in detail in chapter 4. Most had, in fact, more to hide than to display, whether it is what they did—such as their involvement in criminal and sexual activities—or what they lacked (cf. Gable 1997)—that they slept on the floor for want of a bed, skipped meals, or wore trousers with missing buttons and a broken zipper. Indeed, being seen had become preoccupation more than an objective as people generally had more to hide than to display. And, if people had more to hide in this context of vexed expectations and growing socioeconomic inequality, concealment had concomitantly become ever more challenging if only because of growing population density. Youth with *visão* knew how to make the most of the postwar economy and were agile at navigating whichever uncertainty they came across in their everyday lives. They also knew how to do so without compromising respectability.

Disguise and the Politics of Respect

The Cabral Family: Part 1

The Cabrals live on one of the small hills of Liberdade, not too far from the railroad track. Like all the yards in the neighborhood, theirs is kept impeccably tidy. Lush flower beds neatly lined with empty bottles buried neck-deep into the ground deflect attention from the sleeping quarters made of woven reeds and braided palm leaves that are slowly rotting away. The household is headed by Julia, a placid woman in her fifties who works as a civil servant and supplements her meager income by growing vegetables in a nearby field. Of her seven children, only the three younger ones still live with her. Her two sons, both in their early twenties, are struggling to find regular employment despite holding secondary school diplomas. Manuel gets occasional painting contracts whereas Samo does a bit of hustling, selling stolen goods for acquaintances that are not as well connected as he. People in the neighborhood say that he has *maningue visão* (exceptional vision). Both brothers describe themselves as "not doing anything" and can often be found sitting around at home or at one of the neighborhood bars with other young men whose current socioeconomic circumstances offer variations of the same story. Both also devote considerable energy and money chasing after girls from the neighborhood.

Like other young women her age, Sandra, their younger sister, wakes up at the crack of dawn to sweep the yard before heading for school. Domestic chores keep her busy in the afternoons as she is responsible for fetching water at the public tap down the road, cooking, and, when she remembers, watering the plants. Shortly before this snapshot was taken, Sandra started hanging out at a nearby bar to watch the Brazilian soap operas or telenovelas that run in the evenings. Before long, she got into the habit of sleeping out and returning

> home only the following day. At first, her brothers were furious and warned that if she carried on behaving this way, she would never marry. Manuel made it clear that she was jeopardizing her respectability and that of her family. He beat her on several occasions, but this failed to deter his sister. Then, one day, Samo mentioned that their older sister also used to do the same thing, when she was still living with them a few years earlier, but that "at least she had the decency to come back in time to sweep the yard [before anyone could take notice of her absence] and [that] she always cooked a nice breakfast for everyone." He described Sandra, in contrast, as lacking both *visão* and respect (*não tem respeito . . . não tem visão*). In other words, though the brothers openly disapproved of their sister's loose mores, it was Sandra's lack of discretion—along with her selfishness—that truly unnerved them. It was not so much what she did as how she did it that they found deplorable.

Respect is a recurrent trope in the literature on sexuality and intimacy in Africa (Cole 2010; Haram 2005; Heald 1995; Nyanzi et al. 2008), but I think we risk overlooking the complex relationship between respectfulness and respectability if we conflate the two under the notion of respect. While they may be inseparable, respectfulness and respectability are not one and the same. Respectability is normative, used to confer status (cf. Bourgois 1996), and shifting as one usually has to juggle competing notions of respectability such as that conferred by peers, relatives, a professional order, etc. Respectfulness, on the other hand, is performative. In this case, respectfulness is very much reliant on concealment and discretion. In fact, *discretion* and *respect* are commonly used interchangeably. In Liberdade, social worth is largely gauged by how respectful one is and it is not by chance that Samo spoke of respect and *visão* as going hand in hand. "To conceal is respect" (*esconder e respeito*), as Inhambane residents were fond of saying. I will come back later to the waning of male control over female sexuality, but for now I wish to draw attention to the politics of pretense. As this example suggests, *visão* is not only about seeing; it also entails playing on the visions of others. In terms of intimacy, it informs seduction and is also the cunning required to ensure that certain relationships remain under the radar. A closer look at local politics of respect clarifies this emphasis on pretense whereby if one struggles to live up to ideals of respectability, then at least one should be respectful by concealing morally ambiguous pursuits. If Sandra was going to disregard her brothers' authority, she should have at least done so discreetly.

The relationship between concealment and respect is particularly vivid in intimate affairs. "One who conceals does so because he cares" (*quem esconde é porque gosta*), women often comment half-cynically, and those in committed relationships are encouraged to find lovers "far away from home" (*longe de casa*) and do everything not to be discovered. Complaining about her womanizing husband, Benedita further emphasized the importance of discretion when she told me: "Our fathers also used to have lovers, but at least they got them far away, now our husbands go with the 'neighbors.'" In short, respect rests on one's ability to maintain appearances, to play with façades, falsify, and embellish. I show later how it also rests on one's ability to know what not to see, on willful blindness.

An Arsenal of Pretense: Everyday Tricks and Technologies of Concealment

Environments of high visibility often foster the creation of technologies of concealment. Adam Reed (1999), in his work in a Papua New Guinean prison, for example, showed how inmates had a system of hanging blankets in their cells to subvert the panoptic gaze. Thomas Gregor's (1977) research in the Amazon similarly examined ways of dealing with proximity. Gregor used a Goffman-inspired dramaturgical metaphor to show how the spatial setting of the Mehinaku village with its houses built around a central plaza—"as a theatre in the round, one with splendid acoustics and unobstructed seating"—encouraged residents to become "master[s] of stagecraft and the arts of information control" (Gregor 1977: 2). The Mehinaku, like the PNG prisoners, had developed a series of technologies of concealment, tricks and tools designed to render certain things less visible.

In Inhambane, *visão* encapsulates the skills required to successfully toy with other people's visions, to distort what they see or to blind them all together. But to translate into successful concealment, *visão* also depends on technologies of disguise, on a wider arsenal of pretense. One tactic is to move about under the cover of darkness. We saw how Sandra, the young woman introduced in the snapshot above, really should have returned home before sunrise, "in time to sweep the yard." In Inhambane, as in other places in Africa,[26] unlike the day, which is a period of surveillance, constraints, and restrictions, the night comes with freedom, albeit often coupled with some degree of danger. Indeed, night time is commonly associated with thieves, witches, and lovers. There are a few lampposts along rua Branca, Liberdade's main artery, but most of the back alleys (*becos*) that snake around people's houses are pitch black at night, at least until the moon comes out. Walking around at night, one can just make out the silhouette of embraced couples whispering softly. Only people living in the area can safely navigate these dark *becos*. Darkness

26. See, for instance, Davidson 2010; De Boeck 2004; Fouquet 2007; Gable 1997.

saved me more than once from unpleasant and potentially dangerous situations. One evening while having a drink with Kenneth and Jhoker at one of the neighborhood's *baracas*, I found myself having to escape from Mundo, an acquaintance who was very drunk and aggressive. Sensing that the tension was mounting, Kenneth pulled me aside as though he had an issue to discuss with me in private, and we soon found ourselves sheltered by the obscurity. We could still see Mundo and Jhoker, but they could no longer see us. I was struck by how confident Kenneth was of the cover of darkness. We stood in the dark and waited for Jhoker, who was informed of the escape plan by text message. Kenneth whistled softly to signal our position to Jhoker, who responded with the same tune. Being from another neighborhood, Mundo was unlikely to have ever ventured into the back streets after us.[27]

Other technologies of concealment included the black plastic bags used to carry groceries and that kept people guessing what the neighbors would be having for dinner (see also Davidson 2010). Music was also often used, as I indicated earlier, to muffle conversations, and the radio, in particular, served as a technology of aural concealment.[28]

The fences, often over six feet tall, that delimit Liberdade's *becos* were another important component of display/disguise dynamics. Built either out of reeds or woven palm leaves, fences dissuade thieves, lovers, and witches from physically entering a property, and conceal, however imperfectly, possessions, daughters, and wives kept within them. Building a fence is also a grand act of display, not to mention one that suggests that the household possesses something worth hiding. One can, however, also be hiding the fact that one has nothing to hide. The irony is nicely captured by Eric Gable (1997) in his research in Guinea Bissau when a house catches fire, revealing, to the delight of neighbors, that it was actually empty of possessions worth salvaging.

In Inhambane, even poor households without a regular source of income and who live from hand to mouth eventually find the means to build a fence. During a recent trip to Inhambane, I learned that my dear friend Omar had lost his job at the hostel where he had been working for over a decade. Like Omar's life, the reed fence around his property had started falling apart. I saw him and his wife Benvinda regularly during my two-month stay and followed how they managed to scrape some money here and there to get by and feed their two children. Then one day, I arrived at their house to find their son

27. Likewise, a person traveling is likely to prefer leaving and returning while it is still dark to avoid being seen.

28. The Walkman has also been understood to compartmentalize space by creating "a manufactured intimacy" and offering public invisibility (Bull 2004: 113).

weaving a pile of palm leaves to build a new fence. Omar had joked about being able to enter his yard from all sides, but he was clearly determined to rebuild a fence as soon as he could. While fences block the external gaze, they also prevent those inside from seeing what is happening outside, a compelling reminder that display and disguise work together (Taussig 1999).

Young people in Liberdade also relied on linguistic subterfuges—language being "among secrecy's most prominent media" (G. Jones 2014: 57)—to keep certain things hidden, as well as on the careful management of intimate networks. And in recent years they had come to rely heavily on the discretion granted by mobile communication. Indeed, the phone was arguably the most powerful technology of concealment available.

New communication technologies help transcend or even collapse boundaries, especially between the public and the private (Ling and Pedersen 2005),

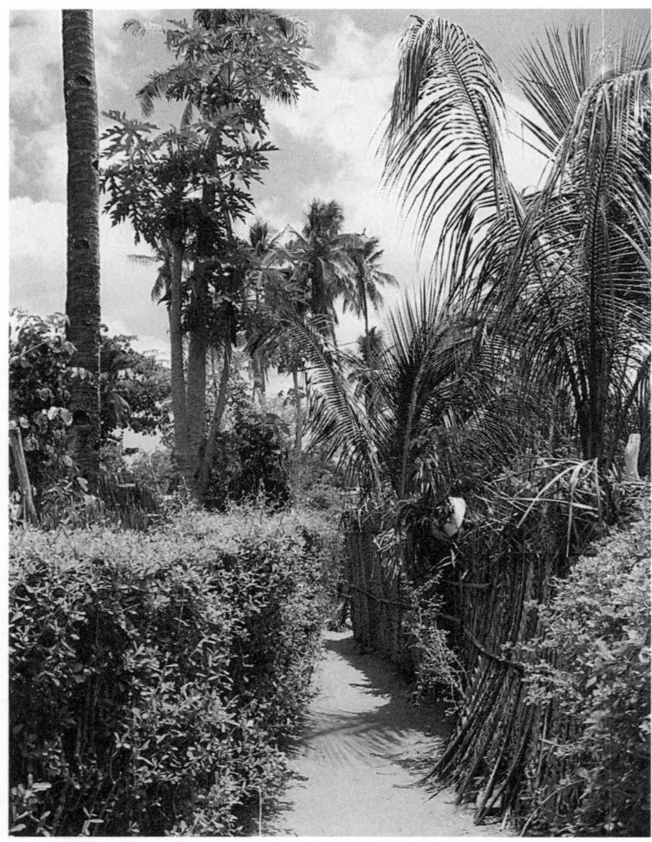

FIGURE 7. The tall fences of Liberdade, Inhambane, 2006. Photo by author.

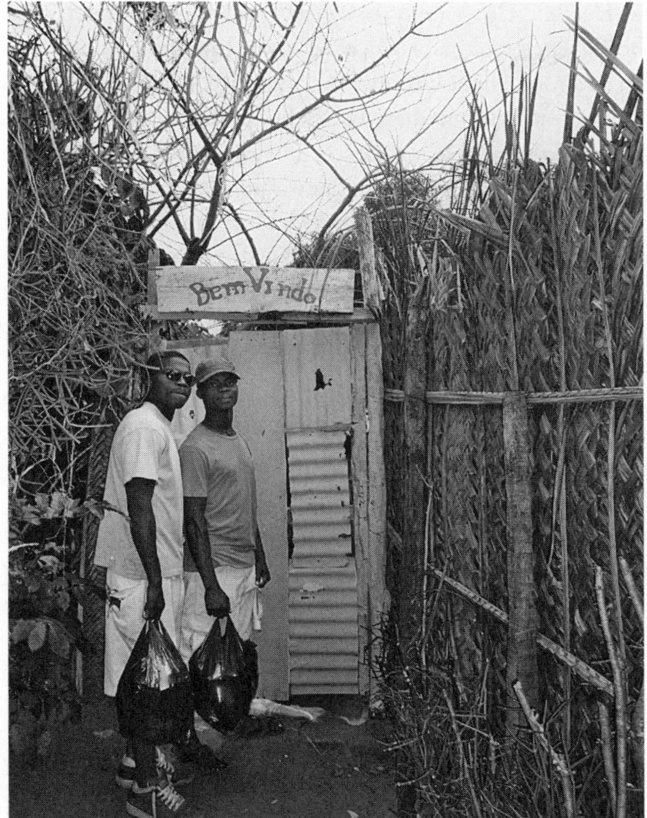

FIGURE 8. Papaito and Kenneth, arriving home with groceries, Inhambane, 2007. Photo by author.

between work and leisure.[29] However, if mobile communication has contributed to the erosion of privacy in some settings (Katz and Aakhus 2002), its impacts on privacy in other contexts seem more ambiguous.[30] In her comparison of the aesthetics of Facebook profiles with American teenagers'

29. The image of collapsed boundaries has also been used to capture the texture of everyday life in contemporary Africa. For example, Filip De Boeck (2004) suggests that, as a result of the ongoing social crisis in Kinshasa, borders that once existed between the invisible and the manifest world, between night and day, as well as between the public and the private sphere, have become increasingly permeable. The same could be argued about Inhambane, where the postwar, postsocialist political economy is also marked by the erosion of boundaries—something that older residents were particularly eager to point out. Authors such as Achilles Mbembe and Janet Roitman (1995) and Harry West (2005) have also written on the collapse of boundaries on the continent.

30. Ito, Okabe, and Matsuda (2005) have addressed the question of privacy in Japan.

bedrooms, Heather Horst (2009) suggests that there is much to gain from understanding Facebook as a place within which people live and interact rather than as a mode of communication. I find it useful to transpose Horst's spatial insight to mobile-mediated sociality and therefore to see the phone not only as enhancing space-time compression, but also as opening new virtual spaces of intimacy, as creating new boundaries and therefore enabling individuals to engage in various pursuits with some degree of discretion. The phone compartmentalizes, however imperfectly, allowing, as Maroon (2006) points out in her research on phone practices in Morocco, "greater opportunity for transgressing moralized social roles" (189).[31] In line with Georg Simmel's (1950) understanding of secrecy engendering a "second world" or level of reality that exists in a dialogue with the manifest world (330), these virtual spaces also have porous boundaries but ones that are no less productive. Unlike the boundaries that Simmel (1950) had in mind, however, the boundaries I am referring to are far more literal, material if you like, even though they are also essentially virtual.[32]

Like I said earlier, no one is entirely oblivious to what happens under the cover of text messaging—the wife knows that her husband cheats on her, the father realizes that his daughter is involved with older men, and the sister is aware that her brother hustles every now and then—but the discretion granted by mobile communication, even in its imperfections, has fundamental implications in a context where growing socioeconomic disparity is coupled with a widening gap between ideals of respectability and actual practices, as it helps preserve and reproduce epistemological uncertainty, and ultimately, some degree of social harmony, provided, of course, that it does not backfire (see chapter 4). By creating new intimate spaces, mobile communication helps mitigate some of the social contradictions characteristic of Mozambique's postwar, postsocialist economy.

Inhambane's historical geographies have translated into particular concerns with display and disguise that are foundational to Mainhambane identity. As noted earlier, *ser visto* (to be seen) has, for many, become a concern rather than an objective. As reflexive residents would say: "A Bitonga prefers to stay at home without money than to be seen in the street doing a low [status]

31. Ito, Okabe, and Matsuda (2005) similarly show how text messaging allows youth in Japan to transcend parental surveillance See also Lin and Tong 2007 for a Hong Kong example and Burrell 2009 for a discussion of internet cafés in Accra as spaces where youth can evade the surveillance of elders.

32. Secrets that are shared by a group of persons, such as initiates of a secret society (Bellman 1984) or craftsmen (Marchand 2009), for example, establish boundaries between members and nonmembers, between those in the know and the rest.

job." Many also identified "Bitonga pride" (*orgulho do Bitongo*) as a key element in the equation—a pride, a particular sensibility, that has its roots in the city's colonial past and that translates into particular "registers of falsification" (Mbembe 2000: 42). Like other young people throughout sub-Saharan Africa and beyond,[33] young adults in Inhambane were, however, not only grappling with the widening rift between the expected and the possible (Vigh 2006: 41), they were also contending with the contradictions between their expansive potential—thanks to widening access to secondary education—and declining employment opportunities (cf. Weiss 2005: 107; see also Mains 2007).

In Liberdade, most young people are seriously committed to completing at least secondary education and many have hopes of pursuing higher education.[34] Like young people all over the world, they have bought into the idea that education offers the most direct and secure route to social mobility,[35] despite also being acutely aware that, in reality, holding a diploma is necessary but far from sufficient. By raising expectations—if I go to school I might be able to secure a job as a civil servant—education has left these young people wanting and waiting. Some have also suggested that alongside the impact of wider global reconfigurations for which neoliberalization is often the preferred shorthand, cultural factors and more specifically the cultural construction of work also play a significant part in shaping experiences of unemployment. According to this culturalist perspective, if young people struggle to eke out a living, it is not simply owing to a dearth of opportunities but also because they covet and feel entitled to specific types of jobs that happen to be

33. There is a large literature that explores the challenges faced by young people across the world. For African examples, see Cole 2010; Hansen 2005; Mains 2012; Peters 2011; Vigh 2006; Weiss 2009. Brison and Dewey 2012, Amit-Talai and Wulff 1995, and Jeffrey 2010 provide examples from other parts of the world.

34. Government figures do, in fact, point to a significant rise in school attendance throughout most of the country from then onward (Mário et al. 2003). In the last decade, three complete secondary schools and a Faculty of Tourism have opened in the city of Inhambane and province-wide secondary school attendance has tripled. Secondary school enrollment rose from 7,504 students in 2000 to 22,150 students in 2005. The data was provided by the Provincial Direction of Education in 2007. In fact, demand is so high that most schools operate on a three-shift system, with some pupils studying in the morning, others in the afternoon, and others still in the evening.

35. A World Bank report (2007: 36) highly commended advances in education, stating that "the capacity of education to deliver social mobility and respectability at the level of the individuals and economic growth at the level of the state has been powerfully propagated around the globe, and education levels have indeed been rising worldwide" (quoted in Amit and Dyck 2012: 13). There are, however, those who argue instead, like Pierre Bourdieu and Michel Foucault did some time ago, that education reproduces inequalities rather than helps transcend them.

particularly scarce. This point is made in passing by Sasha Newell (2012) in his research on youth in Côte d'Ivoire and more forcefully by Daniel Mains (2012) in his work among unemployed men in urban Ethiopia where he identifies shame (*yilunnta*) as accounting for their reluctance to take on low-status jobs. In Inhambane, young people themselves agreed that it was out of pride that they opted to "sit around" rather than engage in menial labor, and graduates in particular saw salaried work in an air-conditioned office as an entitlement (Archambault 2014). I will return to some of the implications of such expectations in the next chapter, which focuses on the petty crime economy.

Before concluding this chapter, I turn to a brief discussion of the politics of display and disguise as played out in local drinking practices. The vignettes that follow nicely illustrate, I think, how disguise and concealment participate not only in hiding things but also in making the world. I add to classic anthropological theories of secrecy, which often fail to fully appreciate the more subtle workings of secrecy in everyday contexts.

In the suburbs, there is much secrecy surrounding alcohol consumption. The more informal drinking venues found scattered across the suburbs tend to be built in such a way that makes it difficult to make out from the outside who is inside drinking. Thatched roofs designed to provide shade and shelter from the rain often go down more than mid-way to the ground. It could be argued that such secrecy is designed to help deflect envy and the pressures of redistribution. For example, Antonio explained that he preferred to drink at a bar in a different neighborhood rather than in Liberdade so as to avoid being seen and hassled by people he knew. Alcohol consumption can, in fact, act as a particularly potent "weapon of exclusion" (Douglas and Isherwood 1979).[36] There is, however, a less instrumental, more performative, rationale also at play here and one that, I argue, taps into the idea of toying with the visions of others. At the time of my fieldwork, it was common for two or three young men to get together to purchase a bottle of the local rum Tipo Tinto, commonly sold in 500 ml plastic bottles. The men would then sit around a table drinking from short glasses while making sure that the rum bottle remained out of sight either on the floor along a leg of the table or in someone's trousers. For onlookers, it would have been plain to see that these men were drinking, but still, drinkers insisted in concealing bottles so as not to attract too much attention. I also saw drinkers get surprisingly upset when too many empty beer bottles would accumulate on the table they were sitting at and waitresses harshly rebuked for this. As will become

36. Akyeampong 1996 and Willis 2002 offer excellent social histories of alcohol in African contexts.

clearer later, regimes of truth in this part of the world are built on what is visible and, in this case, a hidden bottle was like a nonexistent bottle. If these young people acted with such discretion, it was not so much out of shame or greed but rather to "create remoteness" (Sarró 2009)—what others have called *opacity* (Carey 2012)—for others not to know what they were up to. "I don't want everyone to know about my business," Kenneth would remind me almost daily, and especially when he thought I was being too transparent about our research activities. By creating remoteness, Liberdade youth were making claims of authorship over their lives (cf. Jackson 1998: 18–19).

On a research trip to Maputo, Kenneth and I met up with Jhoker, and together we spent an afternoon at one of the small bars near the bus terminal. I ordered a beer and they ordered a Coke to share. The site was rather unusual: two young men sipping Coca-Cola with a woman drinking beer. In reality, however, they were topping up their glasses with whiskey they had purchased earlier on the street and that Jhoker kept hidden in his loose-fitting trousers. The owner was clearly onto them, but as much as she tried to catch them red handed, she was unable to actually see the bottle and therefore had to give my friends the benefit of the doubt. My presence probably also dissuaded her from taking more drastic action. If drinking smuggled alcohol was cheaper, it was also a lot more fun. The two friends had not seen each other since Jhoker had moved to Maputo earlier that year to study meteorology, and by deceiving the bar owner this way they were reminding each other what it meant to be from Liberdade. "Aaaahhh, Mainhambane!" they said. It was all about *visão*.

The concern with being seen conferred to the city of Inhambane a unique feel. Reflecting on the difference in *movimento* (movement, activity) found in district capitals that we visited on the way back from Maputo and which were always much livelier than Inhambane, Kenneth first argued that it was simply a question of money. "There is more money on the National Highway [where many of these towns are located]," he suggested. But upon closer inspection, seeing that most of the people out and about were spectators of the very few actually consuming, he corrected his initial interpretation, and said, with even more confidence, that the difference had to do with Bitonga pride. "If they don't have money to party, they prefer to stay at home."

Scholars interested in secret societies seem to agree that the actual secrets on which these societies are built are secondary to their differentiating force (Murphy 1980). For example, Charles Piot (1993) writes: "Secrecy among the Kabre of Togo seems not so much to hide something real, or exclude access to fixed things (wealth, status), as to set in motion a process—of interpretation, ambiguity, and the quest for hierarchy—and to keep it going" (362). Indeed, control over the circulation of knowledge is also intimately linked to the reproduction of socioeconomic hierarchies (Foucault 1979; Murphy 1980). I would add, however, that the efficacy of secrecy is sometimes far more poetic. Of recent anthropological work on secrecy, Ramon Sarró's (2009) on religious change in Guinea is particularly helpful for the kinds of points I wish to make here. Sarró argues that part of the transformative potential of dissimulation stems from the remoteness it fosters. In other words, concealment helps evade the objectifying gaze of others (see also Davidson 2010: 217) when being known would put one in a position of vulnerability. Individuals lie about where they are going or about where they have been often because the truth would likely land them in trouble or force them to part with something. When it comes to the concealment of intimate relationships, the content of the secret actually matters, and secrecy, in such cases, helps keep people together. As an inebriated young man from Liberdade pithily put it during a monologue in a bar, "Lying exists to facilitate the propagation of the human race!"[37] In other cases, however, they lie, conceal, or falsify essentially so that others will not know what they are up to. They lie, conceal, or falsify because they can. I will come back to this later.

In Liberdade, those with *visão* know how to alleviate hardship in practical terms by activating social networks, by engaging in petty crime, or by exchanging sexual favors for material gain. In this sense, those with *visão* know how to address material insecurity as "a state of limited resources for action" (Whyte 2009: 214). But, as they tap into the new moral economy, those with *visão* also know how to conceal their tracks, how to play with façades. *Visão* is precisely the ability to capitalize on seeing through an opaque social landscape while toying with the visions of others in order to evade not only detection but also objectification.

The arrival of mobile phones may not have radically transformed the lives of Liberdade residents, although many things have changed since the snapshot of the Cabral household was taken. People carry on addressing life's obstacles as they come across them and relying on face-to-face interaction to expand and maintain social networks. But the phone has allowed young people to better cruise through uncertainty. It is through the creation of virtual spaces of intimacy and the redrawing of privacy that this cruise is played out, and ideals imagined, realized, and challenged. In Inhambane, where the "hermeneutic of suspicion" is also "partly the product of a violent past" (Ferme

37. Simmel (1906) himself described secrecy as "one of the greatest accomplishments of humanity" (462).

2001: 7), the ways in which the phone articulates these dynamics reflects, perhaps more than anything, a commitment to peaceful relations. Indeed, the various mobile phone practices discussed throughout the book can be seen as part of a wider arsenal of pretense—albeit by no means an impenetrable one—designed to mitigate conflict and accommodate the demands of intimacy as smoothly as possible.

3

Crime and Carelessness

A handful of people were gathered around the pool table at one of the local *baracas*, the only one in the neighborhood that appeared to be open that evening. Winter was coming to an end but nights were still chilly and Liberdade was awfully quiet, especially on a weeknight like this one. I ordered a beer and sat next to Pascual who discreetly showed me a phone I had never seen before. He whispered that it belonged to Arsenio, the young man sitting next to him, who seemed too drunk to even notice that his phone had gone missing, or as Pascual put it, that his phone had "fallen out of his pocket." We agreed that it would be amusing to see Arsenio's reaction when he eventually noticed. An hour went by but Arsenio showed no sign of cognizance. I suggested that he might only be pretending not to know so as to better investigate the matter, but Pascual assured me that this was highly unlikely; Arsenio simply lacked the *visão* needed to pull off such a feat of pretense. Pascual then decided that, given his obliviousness and profound carelessness (*desleixo*), Arsenio did not even deserve to get his phone back. Pitching his decision as a morally sound one, my friend said that he would instead give the phone to a more deserving young man he knew who was without a phone at the time. Funnily enough, this young man had had his own phone stolen at this very *baraca* a couple of weeks earlier. It had disappeared while he was at the toilet.

Throughout my time in Liberdade, I found myself involved in such incidents in which possessions, some prized more than others, changed hands. My involvement was usually either as a witness, a confident, an observer of the gossip mill, or a victim. Often times, the things lost and found were mobile phones. In fact, such was the frequency of handsets changing hands that phones operated as a form of quasi-currency. Their liquidity on the local market was, as detailed in the coming pages, phenomenal.

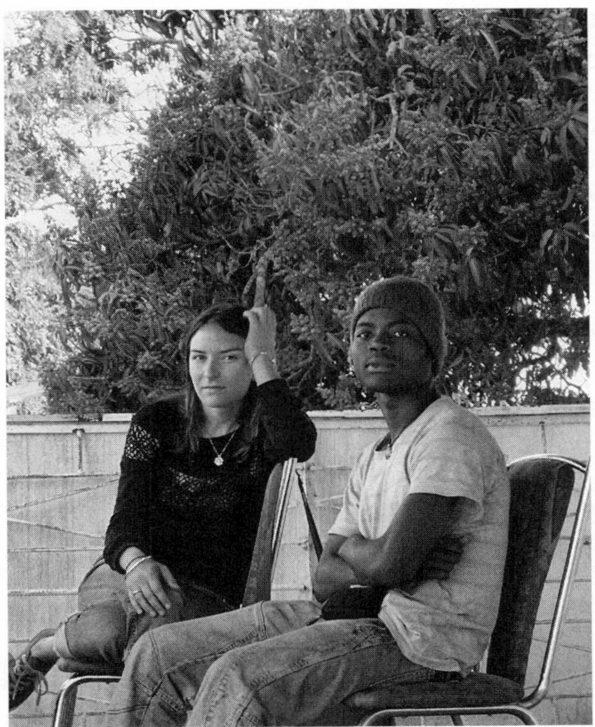

FIGURE 9. Kenneth, my research assistant, Inhambane, 2009. Photo by author.

If phones were tied to crime as coveted objects that were stolen, sold, and resold, many in Inhambane also highlighted the part mobile communication was believed to play in the coordination of assaults and in the liquidation of stolen goods away from the purview of law enforcement authorities, or as was allegedly often the case when the catch was lucrative, with the very assistance of corrupt police officers. Unlike in Europe and North America where the tracking of phone numbers and the investigation of phone records often plays a central role in policing, in Mozambique, phones were more often used to evade detection rather than to disclose criminal activity. At the time, the great majority of mobile phone users relied on pay-as-you-go subscriptions and used unregistered SIM cards.

There are obvious methodological challenges to the study of crime. As Hans Peter Hahn (2012) concludes in an article on mobile communication and highway robberies in Burkina Faso, research on crime can rarely go further than an analysis of local perceptions, rumors, and speculations. I would, however, argue that ethnography has much to offer, provided one can be at the right place at the right time. This is all the more feasible when the thefts in question are a regular occurrence. Although I did not set out to study theft, the workings of the petty crime economy in Inhambane became visible to me through my more general engagement with Liberdade youth. This chapter draws on these experiences and focuses on the part young adults—especially young, secondary school–educated men—play as part-time economic brokers in Liberdade's petty crime economy. Though none of the young people I worked with were full-time criminals in the sense that they relied more on other income-generating activities than on theft to get by, a significant number of the people I knew did dabble in criminal activity every now and then. In fact, several of my male research participants, and a few of my female ones, had spent some time in jail, almost all of them for petty theft, often involving mobile phones. Several of them were in and out of jail during the time I conducted fieldwork. Apart from offering insight into the petty crime economy through the lens of the phone, this chapter also goes deeper into the workings of *visão* by contrasting this essential skill for successful living with carelessness (*desleixo*) and with the need to blow off steam (*desabafar*).

Inhambane, the Land of Good People

Though more than five centuries have passed since Vasco da Gama first came to the area, Inhambane is still known as the "Land of Good People" (*Terra da Boa Gente*).[1] On my first visit to the city in 2001, such an attribute was not only already capitalized on by the tourism industry, it also figured prominently in local identity discourses. When I returned a few years later, I found that a different story was being told. Instead, I heard many people point out that Inhambane was no longer the land of good people. These remarks were part of broader commentaries on the postsocialist, postwar context and its changing moral economy. Among the most vivid indicators of Inhambane having gone "bad" was the rise in criminality. Reflecting on this matter, Kenneth once made the following comment:

> It used to be all about representing your neighborhood. Back then [circa 2000], a guy from Liberdade couldn't go out with a girl from Chalambe [another neighborhood of Inhambane]. We used to fight a lot. . . . But I suppose that back then, we had nothing; nothing to steal from each other. It was before mobile phones.

1. The expression is sometimes translated as "The Land of the Kind People," as *boa* can mean both "kind" and "good."

If Inhambane was no longer the lovely place that it once was, this was, according to young people, because of growing socioeconomic inequality and an ardent yet often times vexed consumerism within which the arrival of mobile phones stood if not as a catalyst, then certainly as a benchmark. It should be noted, however, that when Inhambane residents talked about Inhambane no longer being the land of good people, they were not only commenting on rising criminality, but were also hinting at a wider moral degradation manifest in the reconfiguration of the intimate economy, a topic I cover at length in the following chapters. In fact, alongside an increase in attacks on personal property came a rising incidence of crimes of passion, with adultery singled out as the main motive for homicide.

When phones first started trickling into the area just after the turn of the millennium, phone muggings were reportedly frequent and various locations in and around the city became renowned for assaults. As penetration increased, phone theft turned more surreptitious as well as more mainstream. In the survey I conducted among secondary school students, over 30 percent of phone owners reported having a phone stolen at least once in their lives.[2] Instead of being stolen in holdups, phones were increasingly swiped discreetly through pick-pocketing and opportunism, of which Arsenio's lost phone offers a good example. Women were also, on the whole, more vulnerable to phone theft than men.[3]

It was often late at night that the most ethnographically interesting events in Liberdade unfolded, and it was usually during my morning meanderings in the neighborhood that I heard the most exciting stories. It was then, when they were more likely to be on their own, that my young companions would take the opportunity to confide in me, to *desabafar* (unload, blow off steam), and fill me in on what they had been up to the night before.

There is an interesting parallel to draw between *desabafar* and *desleixo* (carelessness). *Desabafar*, or the act of taking something off one's chest, is recognized as a fundamental need with therapeutic benefits. It is, however, a highly risky indulgence, given, as detailed earlier, the importance of concealment and the dangers of being known. In turn, offering an attentive and empathic ear to those in need of *desabafar* is a moral obligation. *Desabafar*, as an inescapable need, is particularly apposite in my discussion on visibility and pretense as it suggests the expression of sentiments that are genuine and

2. Or, as it were, within the last three years, as the great majority of the young people in my research had acquired their first handset in 2004.

3. This is according to the survey I conducted in 2007.

involves momentarily putting one's guard down and saying things as they are, usually to voice dissatisfaction with something or someone. As implied in the formal meaning of the word, which is "to uncover" or "to reveal," to *desabafar*, like *desleixo*, involves the suspension of pretense. Doing so entails opening up and putting oneself in a position of vulnerability. Though what is said may sometimes be calculated and strategically voiced in the hope of reaching some sort of objective, in general to *desabafar* is understood to operate according to a different logic.

What makes petty crime ethnographically interesting in a discussion on the politics of pretense is how it takes us to the interface between hyperawareness and momentary carelessness. A look into theft brings to light how visibility and invisibility work together in young people's efforts to cruise through uncertainty and helps illustrate how *visão* rests on the ability to see while playing on the visions of others.

It is tempting to conceptualize the criminal economy in opposition to the formal economy, to explain the importance of one by the limited opportunities available in the other, and to fall back on a Durkheimian view of crime as a symptom of anomie that manifests itself in periods of transition. Petty crime has in fact been understood not as a way of life but rather as a tactic, among others, to address needs and aspirations unfulfilled by other means (Kyed 2008; Mbembe and Roitman 1995: 337). A look into the workings of petty crime, then, raises questions central to the Africanist literature on youth around production, consumption, and purpose. In a postwar, postsocialist context where consumption is increasingly overshadowing production in the realization of self, crime often offers an appealing alternative. Like in other settings, however, illegality is imbricated within spheres of legality in ways that can simultaneously weaken and strengthen more formal structures (Humphrey 1999). Even more importantly, by understanding crime solely as a symptom of something else, we risk overlooking how crime also operates as a socially meaningful practice.

Anthropology was initially interested in crime as a moral rather than as a legal issue.[4] In the small-scale societies studied by anthropologists throughout much of the twentieth century, crime was commonly apprehended as a moral violation and offenders were often punished through shaming (Schneider and Schneider 2008: 354). It was in *Crime and Custom in Savage Society* that

4. Anthropological interest then shifted toward plural or dual legalism in a world where customary law came to operate in parallel first to colonial law and later to the laws of independent nation-states.

Malinowski ([1926] 1966) first set up his theory of reciprocity as "the basis of social structure" (46), which was to have a lasting influence in social anthropology. Malinowski's thesis was meant as a corrective to the pervasive view at the time that "primitive man" followed custom blindly or essentially because of fear of spiritual reprisal. Instead, Malinowski suggested that the main principle behind law and order was reciprocity. Through his focus on reciprocity and his emphasis on the importance of methodological skepticism,[5] Malinowski recognized self-interest and personal ambition as important social forces.

Meanwhile, although pioneering ethnographies such as Margaret Mead's (1928) *Coming of Age in Samoa* transformed adolescence[6] and its related fields (socialization, development, initiation, intergenerational relations, marriage, etc.) into classic topics of anthropological inquiry, youth remained a transitional and teleological phase modeled by adults and oriented toward adulthood (Bastian 2000; Bucholtz 2002: 525–32). Only since the mid-1990s (Amit-Talai and Wulff 1995) are youth portrayed not merely as adults *en devenir* but also, and perhaps more importantly, as social actors capable of shaping society (Bucholtz 2002: 525–26; Durham 2004). Recent studies have gone beyond underscoring cross-cultural variability of childhood and adolescence and owe much to research set in sub-Saharan Africa, where youth ambivalence, epitomized by the child-soldier (Stephens 1995: 19), has inspired a rethinking of youth agency.[7] Much contemporary work in the anthropology of youth has grappled with the question of youth ambivalence (Abbink 2004; De Boeck and Honwana 2005). This chapter delves into the ambivalent position youth occupy in such an uncertain environment and offers insight into the ways in which they wrestle with inequality. Criminality, like unemployment, cannot, however, be reduced to an epiphenomenon of neoliberalization. It deserves, instead, to be situated in relation to historical specificities and cultural preoccupations. I therefore prefer to approach petty crime not as an index of uncertainty—even if it arguably is also partly that—but rather as a social practice with its own productive and destructive consequences.

5. Malinowski insisted that anthropologists had to go beyond narratives—beyond what people say they do—to understand social life.

6. As Mary Bucholtz (2002) points out, the etymology of adolescence refers to the idea of a transitional phase: "*Adultum* is the past participle of the Latin verb *adolescere* 'to grow (up).' The senses of growth, transition, and incompleteness are therefore historically embedded in *adolescent*, while *adult* indicates both completion and completeness" (532).

7. See also Gable 2000; Rasmussen 2000.

Pride and Entrepreneurialism

When young people in Inhambane dabbled in petty crime, they were seizing opportunities and taking concrete measures to address their exclusion from the labor economy and to craft successful lives, especially as the benefits were generally higher than those of regular engagement in menial labor (cf. Bourgois 2002). If unemployment is often cast as a driving force behind criminality, the part sociocultural factors play in the equation is certainly less understood. Here I wish to revisit the articulation between pride and visibility introduced in the previous chapter in my discussion of the political economy of display and disguise.

In his research on street vendors in Maputo, Agadjanian (2005) explains that in an attempt to preserve their reputation, their "social manhood," young men who rely on street vending—not a particularly glamorous occupation—"choose their peddling routes so as to minimize the chances of embarrassing encounters with their peers and girlfriends" (262). Some of these vendors congregate at the Junta, the bus terminal for coaches and minibuses heading north from the capital. One morning aboard a minibus bound for Inhambane, the passenger sitting next to me greeted a vendor he recognized from Inhambane where, I gathered, both men were from. The vendor was visibly embarrassed to have been discovered, despite the passenger's clumsy attempts at persuading him that there was nothing to be ashamed of. The passenger then attempted to purchase one of the seller's pens, but he only had a large bank note. The seller looked mortified as he rummaged through his pockets for change before darting off. I often heard young people praise those who engaged in menial work, probably as often as I heard them admit that they would rather go hungry than do so themselves.

Inhambane is much smaller than Maputo. There is no place to hide, no alternative peddling route. Unlike the unemployed young men in Kenya who have been described as "tarmacking" as they walk up and down the road in search of work (Prince 2006), Liberdade youth were reluctant to render their predicament visible for all to see. In contrast also to other parts of the country where almost everyone runs some sort of small business, the low level of entrepreneurialism in Liberdade was remarkable. Less than a handful of the young men I worked with were involved in some form of commercial activity. In fact, young women were more likely than their male peers to have their own small businesses braiding hair or selling something such as popsicles, roasted corn, or alcohol. Menial labor was seen as reserved for recent immigrants. Gitonga speakers, for their part, argued that they were unable to start their own businesses for want of capital. But the reflexive ones also recognized that it was essentially "Bitonga pride" that quelled entrepreneurial initiative. Reflecting on his experience in East Germany where he was sent to study in the early 1980s, a middle-aged policeman explained that what had struck him the most during his time overseas was seeing students working as street-sweepers. He was convinced that no young person in Inhambane would agree to sweep the streets, even if their life depended on it. They had a sense of entitlement, he continued, and preferred to wait for government jobs, even if this meant going hungry ("*pasar mal de fome*") in the meantime. If "waithood" (Honwana 2012) was more a metaphor than an accurate rendition of these young people's reality, it nonetheless shaped how time was experienced and allocated. Having studied not only made stasis more painful, as graduates were expecting, and felt they deserved, some level of social mobility; it also justified it. Instead of having to make a living, "to search for life" (*ir à procura da vida*), whichever way possible, graduates could justifiably "wait" until an acceptable opportunity presented itself.

I once hired a young woman called Balsa to act as a part-time research assistant. On our first day out, we headed to one of the city's secondary markets to talk to the young people working there. Balsa, who at the time was living with her two daughters and ailing mother in a modest reed house in Liberdade, made it clear that she was uncomfortable with the agenda I had set for the day. "I have a flaw," she admitted. "I don't like to be seen in lowly places." I had had a string of unsuccessful attempts at enlisting the help of research assistants who had all showed contempt for the "marginalized youth" I was interested in knowing better, despite themselves fitting into this loosely defined social category.

One of the ways of avoiding embarrassment, aside from "staying at home," was to find work abroad. One man who had work experience in South Africa explained: "Going to work in South Africa is fine because over there, no one can see what I'm actually doing. I just come back with money and nice things and that's fine." That said, and a few exceptions aside, labor migration remained very much the preserve of rural men[8] and very few urban secondary-school educated youths saw it as a desirable option. In fact, I found nothing even remotely comparable to the "culture of exile" that Charles Piot (2010) encountered in West Africa. Southern Mozambique's history is one marked by migrant labor to South Africa, but, as mentioned earlier, migrants usually

8. Several young women from Liberdade had recently been to South Africa to work in hair salons because of their expertise in braiding hair. They had returned with enticing tales, as well as money and nice clothes, and were convincing others to join them. This said, the great majority of migrants (93.1 percent) are men (de Vletter 2006: 7).

remain invested in the communities they leave behind, intend to retire in Mozambique, and also return more or less regularly for holidays and special occasions. In fact, only a small percentage of Mozambicans actually immigrate to other countries. Portugal attracts scores of immigrants from its former colonies but although Mozambique is Portugal's most populated former colony in Africa, followed closely by Angola, Mozambicans only constitute 25 percent of Portugal's Luso-African immigrant community.[9]

Another option was to sit tight until a desirable job eventually came along and engage in a bit of discreet hustling in the meantime. Indeed, Bitonga pride did not so much curb entrepreneurialism as a whole; rather it curtailed a particular kind of entrepreneurialism. A closer look at young people's livelihoods revealed complex, often mercurial, entrepreneurialism in the sexual and/or criminal economies. What I found was that mobile communication had opened up an "alternative peddling route" (cf. Agadjanian 2005), so to speak—one that allowed individuals to engage in all sorts of activities while evading public scrutiny. So long as they did not get caught, of course. But even then, there was always some degree of uncertainty surrounding guilt, as it was not unusual for young men to be framed or wrongly accused. I will return to the workings of the sexual economy in chapter 5.

The "Little Bandits" of Liberdade

I had been living in Liberdade for just over a year when a group of young people from the neighborhood decided to put together a lunch for the elderly.[10] During the month leading up to the event, the organizers managed to collect sufficient contributions from residents as well as from various local sponsors to cater for over a hundred people. The word quickly got around and, on the day, diners—the majority of which were elderly women—came from near and far, some even arriving at dawn to secure a spot in the shade. The organizers had to borrow plates and cutlery from several members of the community, purchase food on a budget, and wake up before the crack of dawn to cook in large pots over open fires. It was a lot like a wedding except that in this case the men did much of the cooking. And unlike at weddings, where the atmosphere is almost inevitably dampened by critical commentary about the food lacking salt or the alcohol running out too quickly, only

9. According to the 2011 census in Portugal, 69,430 Mozambicans were living in Portugal at the time (http://www.ine.pt/xportal/xmain?xpgid=ine_main&xpid=INE, accessed December 9, 2013).

10. The lunch was held on October 1, 2007.

praise could be heard. One of the event's highlights was the improvised performances of some of the male organizers who sang and danced to the delight of the elderly diners. Overall, everyone present agreed that the event was a tremendous success.

The main objective was "political," I was told by João, one of the masterminds behind the event, who later explained that the lunch was "meant to show everyone that there [were] youth with *visão* in Liberdade." In my view, this community engagement and solidarity stood in contrast to another side I also knew of my Liberdade friends—one of young men and women who often stole each other's clothes, shoes, phones, not to mention each other's girlfriends and boyfriends. Later that day, I expressed my surprise to Kenneth and asked him how a neighborhood where crime was so rife could also be one with such voluntarism. Where I saw paradox, Kenneth saw limpid coherence, as he responded, "Bandits have exceptional *visão*; they are clever!" Not only was my understanding of conflicting moralities fraught, it also overlooked the workings of similar cultural logics in both spheres where *visão* emerged as a set of transferable skills that could be used for constructive and destructive ends.

In Liberdade, falling victim to crime was experienced not only as a proprietary offense, but also as an attack on one's dignity, especially when the assault involved swindling rather than brute force. Victims sometimes received sympathy but most of the time, like in Arsenio's case, falling prey to theft, especially phone theft, was seen as a sign of carelessness (*desleixo*); it spoke of a lapse in attention that stood in sharp contrast to the state of hyperawareness required of the successful urban dweller. Blame, then, was usually directed toward the victim. And when the victim was wealthy, onlookers were also likely to find comfort in a sense of retributive justice.[11]

Having a phone stolen is a pain: you need to buy a new phone, buy a new SIM card,[12] reacquire contacts—though many kept handwritten backups of important numbers—and disseminate your new number.[13] There was, therefore,

11. Burrell (2010) reports a similar attitude in her study of international Internet scamming in Ghana where scamming is rationalized as "vigilante justice" to address global inequalities between rich "thieves" and poor Africans (23).

12. Stories abound of criminals in Maputo who hand over SIM cards to their victims before running off with their phones.

13. Although phone numbers could be recuperated for a small fee paid to the network provider, most young people I knew who had fallen prey to phone theft felt the cost was prohibitive and preferred to buy a new SIM card, which came with a new phone number. At the time, one also had to factor in the trip to Maxixe, the closest place where this service was available. All three mobile operators have since opened stores in Inhambane itself.

much pride in holding on to an original number. Unlike his brother who would boast about still using his very first phone number, Papaito never lasted very long with a new phone. On one occasion, Papaito had been drinking at a local *baraca* and, although his house was only a short walk away, he was so drunk that he had passed out in an alley. By the time he regained consciousness shortly after sunrise, someone had emptied his pockets and even taken off with his flip-flops. His older sister laughed as Papaito told us the story the following day. She suggested that it was probably one of his friends who had emptied his pockets. "Had Papaito woken up," she said, "the 'friend' would have said that he was only trying to help [by safekeeping his things] before someone else robbed him."[14] For the young men who spent time out and about, the street, or the alley as it were, was a space in which trust was scarce and alliances fluctuated constantly. It was at times a space of trouble, at times a space of opportunities.

The distinction between victim and perpetrator, like the dichotomy between lender and borrower (cf. James 2012), for instance, is one that risks overlooking the complex relationships that underpin these local economies, not to mention the ways in which the same individual can alternate more or less regularly between the two. That said, if the same person can be a victim one day and a perpetrator the next, these experiences are, of course, qualitatively different and, as the distinction between *visão* and *desleixo* (carelessness) suggests, it would be wrong to see them as reflecting a continuum of awareness. One's carelessness might become another's opportunity but, whereas *visão* is something that one cultivates, an essence almost that becomes part of a person's very being, *desleixo* is a momentary hiatus, albeit one that may have dire and lasting consequences.

Consider the following two examples from my fieldwork, which offer convergent assessments of the social value of petty crime. During the summer of 2012, I was staying at a hostel in the city center of Inhambane. Among the other guest-residents was a Dutch filmmaker who was making a film based on local reactions to a crying bride in different countries including Japan, Russia, and Mozambique. One Friday evening, I was heading to Liberdade to meet a friend for a drink and I invited him to come along. Knowing that he had never been to the suburbs before, and feeling responsible for his welfare, I casually warned him, on the way there, to mind his pockets. To my surprise, the filmmaker did not take kindly to my words of advice. He responded that he had

14. I did witness instances in which individuals safe-kept a drunken friend's things until the following day and others in which phones were taken from inebriated men who had let their guard down.

been in the "land of good people" for over a month already and that nothing bad had ever happened to him. He even accused me, as an anthropologist, of fabricating and propagating the criminalization of Mozambican youth. We eventually arrived at our destination where we continued a rather heated debate that turned to ethics. The argument was in English and the meaning lost on the other patrons of the *baraca*. The filmmaker eventually caught a taxi back into town and left me behind with a small group of young men who were curious to know what the argument was all about. I willingly summarized the filmmaker's accusations and also tried to get a conversation going around the criminalization of African youth, but no one seemed interested in elaborating on the topic. The men agreed, however, that I had done the right thing in alerting the filmmaker to mind his pockets. One said, and the others agreed, "Ahh . . . here there are many bandits!" He added that newcomers to the *baraca* that evening had been informed that "the white man" was with me and told to leave him alone. So I rest my case, dear filmmaker! As the following example also nicely illustrates, young people's cunning—more than their involvement in petty crime itself—can, indeed, be seen as something to be proud of.

The following Friday, I bumped into Sale, a young man from Nampula, a province in the North of Mozambique, that I had met in Inhambane over a decade earlier when he was the gopher at the place I was staying. He was only passing through, on his way to Maputo, as he waited for his truck to be serviced in Maxixe, the city on the National Highway on the other side of the bay. Sale was itching to have a drink so I suggested we go to Liberdade where I knew there would be more activity than in the city center. On the way there, Sale filled me in on what he had been up to since the last time we had seen each other. He told me stories about diamond smuggling in Angola and about how he had been involved in terrible things that he did not even have words for. "I'm not like your friends here; they're just *pequenos bandidos* [small time bandits]," he told me, as he tried hard to persuade me that he was a first-class criminal. When we arrived in Liberdade, it was a little awkward as Sale refused to socialize with anyone and carried on with his disparaging remarks. I ended up leaving without him. The next morning when I caught up with Sale again, he was ashamed to admit that "they"—the "little bandits"—had robbed him blind. Later in Liberdade, no one I spoke to knew for sure who had taken Sale's stuff, but the deed was claimed as a collective achievement and everyone thought it was well deserved given his pretentious manners. "We showed him," one of them said succinctly, "you don't mess with people from Liberdade." To be honest, I felt rather proud of the little bandits of Liberdade.

Despite their different outcomes, both events highlight the part criminality plays in young people's attempts to craft fulfilling lives. In a sense, the filmmaker was right to challenge the idea that my companions were potential thieves. At the end of the day, it was arguably youth entrapment more than criminal intent that was to blame for their dabbling in crime. This was, in fact, how some in Liberdade saw it. For example, after spending time in jail for stealing bags from the car of South African tourists, Jhoker told me:

> These are youth games. Sometimes life pushes us in certain places, owing to certain problems. . . . Here it's very difficult, you know, sometimes it's money we need. So we violate because we see that "I can't believe it, how can I just endure these frustrations when there's another way." Myself, I ended up falling in a temptation.

But the filmmaker was wrong to assume that they would consider such a label as purely derogatory. It was not that theft was glorified in itself—in fact, theft was commonly described as ugly—but rather that it was a manifestation of *visão*, even sometimes at the level of the neighborhood, which, when deployed in the right circumstances, was deemed as something to be proud of. If the power to define the criteria of illegal predation tends to rest in the hands of the powerful (Schneider and Schneider 2008)—and there are particularly appalling examples of criminalization, such as when "bad laws" force individuals and sometimes entire communities into illegality (Vasquez-Leon 1999)—the less powerful nonetheless also play a role in defining the legitimacy of certain illegal practices (Larkin 2008; Roitman 2005), along with their social worth. Petty crime brings unequal social relations into relief, while also serving as a mode of redistribution *and* as an important medium of self-fashioning.

Levelling Inequality: The Phone as Quasi-Currency

I showed, in the chapter on the political economy of display and disguise, how growing disparity was experienced as particularly painful given the legacy of socialism that emphasized equality and the abrupt transition from a wartime economy of extreme scarcity to one characterized by a sudden influx of modern consumer goods that remained beyond the reach of the majority. In Inhambane in particular, the tourism industry significantly contributed to the provisioning of the local market and has had a very tangible impact on the mobile phone landscape and on the availability of handsets. Indeed, a number of phones make their way to the city in the pockets and on the dashboards of foreigners before being inadvertently "left behind." These phones are then inserted into the pool of goods that neighborhood petty crime stirs up some more.

Stolen handsets are relatively easy to liquidate, so long as they are not set on an incomprehensible foreign language. Handsets injected this way into the local economy then circulate further via various channels of gifting, exchange, and theft. Phones are also sold for upgrade or when a pressing need for cash arises such as to pay for bus fare when stranded in another city, or to cover unexpected medical bills. Indeed, phones are also valued for their store value and could be seen as a form investment. As Zito put it to me, "I bought a fancy phone and everyone accused me of being a show-off but what I did, I did with a plan in mind as I knew that if I needed cash I could simply sell my phone." Money spent on handsets is money protected from the pressures of redistribution as well as from one's own indulgence (Archambault 2014). But phones clearly offer a particularly high-risk form of investment.

If mobile money transfer services such as Kenya's famous mPesa have attracted much attention (Duncombe 2012), far less is known about the more informal forms of mobile exchanges in Africa that take place every day. There are in fact several creative forms that mobile money takes, from airtime transfers as gifts, to the purchase of handsets and airtime as store value, to the circulation of handsets in circumstances such as those detailed above. The introduction of mobile phones has given young people something to trade in and it is for this reason that I find it useful to think of handsets as quasi-currency, as portable yet not entirely liquid things, that circulate and are used, not unlike money, to gain access to other things. The circulation of mobile phones also deserves to be situated within broader systems of exchange according to which there is also a tacit understanding that things are meant to circulate and that ownership is rarely, if ever, a permanent affair.[15]

Some of the young men in Liberdade recalled a time, circa 2000, when they would go to the beach every now and then and follow tourists around in the hopes of getting their hands on valuables such as sunglasses, a pair of nice sandals, or Billabong swimming trunks that were popular at the time. They used the verb *to harvest* (*colher*) to describe this activity, pointing to how easy it was to take things from these unsuspecting tourists who had, after all, bought into the idea that they were vacationing in the land of good people. Since then, the tourism industry has seen a massive boom with lodges and self-catering chalets now dominating what was, until recently, a practically

15. When my house was robbed my neighbors tried to console me by saying that my things were not meant to remain with me forever. Needless to say, this did not make me feel any better. Everyone agreed that the thieves were acquaintances of mine and that I was foolish to have trusted anyone, especially young men. The police officer who investigated my case, himself a Mozambican, advised me to no longer socialize with "Blacks"!

untouched coastal landscape. Inhambane's pristine coastline and its relative proximity to Maputo[16] and to the South African border make it an increasingly popular destination, especially during South African holidays.[17] In anticipation of the 2010 World Cup, which was held in neighboring South Africa, I heard one person say, "I can't wait for the World Cup: more tourists means more phones and iPods for us!" Were it not for petty crime, tourism's direct contributions to the local economy would be marginal as the great majority of the resort owners are South Africans and payments are often made directly into South African accounts. Moreover, most of the accommodation on the Inhambane beaches is self-catering and caters mostly for South Africans who commonly come into Mozambique with most of their consumables (food, drink, beer, toilet paper, etc.).[18] The city itself struggles to attract tourists, who tend to head directly to the nearby beaches and beach resorts situated approximately twenty-five kilometers from the city center. Visitors cannot, however, completely bypass the city, as the road that leads from the National Highway to Tofo and Bara beaches, the main beach resorts in the area, runs across town. A number of bars also line the road and patrons can sometimes be heard making cynical comments as they watch tourists driving in. During the high season, the sight can be rather stupefying: lines of sport utility vehicles (SUVs) filled with people and things to maximum capacity and often towing anything from trailers to boats to quad bikes. Because they usually keep their windows up to enjoy the air-conditioning, this only accentuates the deeply rooted disjuncture that still exists between these two parallel universes—between a rich South Africa and a crippled, albeit growing, Mozambique—with interconnected histories. City residents are therefore in frequent yet rather superficial contact with foreigners, and consequently have a completely different understanding of what tourism entails to those living near the resorts and who have more opportunities to develop intimate relationships with foreigners. They do, however, have ample reasons to

16. The city of Inhambane is situated 492 kilometers from Maputo and most of the road connecting the two has been rehabilitated. According to the Provincial Direction of Tourism of Inhambane, an estimated 10,000 tourists visited the province of Inhambane every year during the Christmas/New Year holidays.

17. Of the province's twelve districts, eight have access to the coast, which stretches from the Limpopo and the Save rivers.

18. South African friends of mine came to Inhambane while I was conducting fieldwork, and together we went on a five-day camping trip. They had brought food and drinks, and the only thing I remember them buying during their stay was bread, in addition to a dinner at a touristic restaurant.

fantasize about their air-conditioned "cars filled with riches." Young people's relative ease of access to phones and other digital devices would be considerably stunted were the constant supply of handsets brought in from abroad to dry up.

On a scorching December afternoon, I stood with Lulu by the side of the road looking at the seemingly endless queue of SUVs entering the city. Lulu said: "The tourists that are coming in now are going back without clothes!" It was one week before Christmas and two weeks before Lulu, a young man from Liberdade who had already been locked away once before despite being the son of an influential police officer, was apprehended for suspected robbery. Lulu's arrest prompted interesting comments from other young men in the neighborhood. One said, "We were also little bandits in our time but even then we knew that life is not done that way," and another added, "Lulu doesn't want to learn, but we've learned already." Indeed, no one I worked with saw crime as a permanent career option, as an alternative to a more conventional livelihood. Instead, dabbling in crime was meant to be only a phase, something to do "in the meantime" (*por enquanto*), until a proper job came along. In other words, no one saw crime as a way of life. The commentaries also hinted at the difference between narrow-sighted and farsighted individuals and, ultimately, at failure to dupe others. What many had learned was in fact how not to get caught.

In this economy, the phone was not only an object of exchange, it was also used to set other things in motion. Young men who were involved in petty crime had developed networks of potential buyers among the employed residents of the neighborhood. On several occasions, I witnessed how these young men would start up conversations with older state employees—or anyone that seemed to be salaried—in bars where intergenerational encounters were common, and how they would find a motive to exchange phone numbers. Professionals, including teachers, doctors, and police officers, were generally very open about giving out their private phone numbers. Some young men acted as brokers and would help find buyers for those with something to sell but who were not as well connected as they were. To be activated, these informal channels relied on discretion, along with astute network management skills.

Throughout my stay in Inhambane, I often heard discussions about crime and, more specifically, about theft. "People here, how they like to steal!" was a comment I heard on numerous occasions. In some cases, young men were harshly criticized by older residents for what they described as young people's "materialism." Recalling his youth, a father of six said:

> I grew up on sugar cane and bananas. Youth nowadays prefer to sell the bananas instead of eating them. It's all about money. . . . Back then young men would collect a can of cashew nuts and sell it to an Indian merchant in order to buy a nice pair of trousers. . . . Now they own trousers worth 500 MZN but they don't even work!

Along similar lines, another father explained that his son had brought home a new flashy handset but that he felt unable to inquire about its origin. I did not, however, come across the same level of exasperation with theft as reported elsewhere in Africa (Pratten 2008b) and beyond (cf. Caldeira 2000; Goldstein 2003).[19]

There was a notable contrast between the rise in vigilantism reported in many places and the near absence of vigilantism in Inhambane, despite the city's relatively high petty criminality. The city did have a growing private security industry, working alongside informal unarmed guards as well as local-level structures of surveillance with popular police elements living within the population. But there was nothing comparable with, for example, the vigilantism described by Steffen Jensen (2008) in Nkomazi in neighboring South Africa, where influential individuals, customary authorities, and youth have taken it upon themselves to maintain security often through violent means. It is true that crime in Inhambane was generally far less violent than in South Africa, but I think that there were other factors at play in shaping local responses to petty crime.

The Liberdade residents who were renowned thieves were identified as such but were not necessarily marginalized; instead, other residents maneuvered around them with a combination of reverence and fear (cf. Roitman 2005). Their contributions to the local economy were, in fact, implicitly acknowledged through tolerance (see also Schneider and Schneider 2008: 362). Like piracy, which Brian Larkin (2008: 14) has described as a "mode of infrastructure that facilitates the circulation of cultural goods," petty crime helped democratize access to certain goods by putting them into circulation and, at the same time, by bringing prices down. In this economy, today's victims were often tomorrow's winners. The circulation of phones and other goods

19. Cases of lynching have become common in other Mozambican cities like Maputo, Beira, and Chimoio (Serra 2008). In 2013, panic settled in the suburbs of Maputo as a group of armed men known as the *engomadores*, in reference to their use of irons to mutilate their victims, were believed to be roaming the streets at night, targeting houses. The government soon dismissed the very existence of this group and blamed the panic on rumors.

not only altered the value of the objects themselves, it also participated in shaping sociality and in rendering friendships highly unstable and uncertain.

'MOBILE' PHONES AND THE CYCLE OF FORTUNE AND MISFORTUNE

The following story, starring Papaito in the leading role, took place over a two-month period and involved at least a handful of mobile phones.[20] I have included it here to give a sense of the workings of the petty crime economy, of the fluctuating nature of alliances, and of the part mobile phones play in these dynamics as a quasi-currency that often pits Liberdade residents against each other.

> School was finally out for summer and Dama do Bling, one of Mozambique's leading female rappers, was giving a concert to mark the start of the holidays. On their way to the show, which was held at the training ground in the city center, Papaito and Lulu ran into Abibo, another young man from the same area, who was having a beer at one of the local *baracas*. Abibo had recently started working as a money collector in a minibus taxi and was one of the few who could afford to drink beer—the preferred evening drink during the hot summer months—instead of the considerably cheaper local rum. Abibo bought a round of beers, but Papaito and Lulu took offense as they saw his gesture as a pretentious display of resources. They took a few sips and then headed off, leaving Abibo behind with a set of half-emptied glasses. The tension between sharing and showing off was later nicely captured by another young man who said: "Here, if you buy [someone a] beer, it's a problem, but if you don't, it's also a problem."
>
> Once at the concert, the two men were approached by a man selling a mobile phone he had just stolen inside the concert venue. The asking price was 350 MZN (about $20). Papaito, who had recently lost his phone on a night out, as detailed earlier, offered 300 MZN but meanwhile, Abibo, who had caught up with them, said that he would happily pay 400 for the phone. Insulted for the second time that evening, Lulu and Papaito gave Abibo a good beating. According to Papaito, who recounted the story to me the following day, Abibo needed to be educated on how to handle money. Meanwhile Papaito remained without a phone.
>
> The following Friday night, a number of young men from the neighborhood, including the three protagonists, and I were having drinks in the

20. The events started on October 21, 2007. Some I witnessed firsthand and others were recounted to me.

neighborhood before heading to the full moon party held at Tofo Beach.[21] All of a sudden Abibo, the one who had been beaten up a week earlier, started panicking: his wallet and phone had disappeared. Others tried to investigate the matter until someone eventually found the missing wallet near the toilets. It was empty. We all headed to the party as planned despite the turn of events. The night was pleasant and surprisingly uneventful, at least for most of the Liberdade posse. Lulu, on the other hand, was not as fortunate. He was caught up by the security guard after trying to leave without settling his tab and was beaten quite severely. Lulu, who was carrying two phones at the time, came back home with none.

The following day, Papaito told me that he was the one who had taken Abibo's wallet and that he had spent the money at the party. He added, once more, that he had done so to educate Abibo, "to show him that he needs to treat people with respect." He also said that he would probably end up refunding him one day. "Life is a cycle," he explained. "Now Abibo is paying for the harm he did. One day it'll be my turn. And, when that day comes, I'll know that I'm paying!" It was, in fact, not going to take long for the cycle to come full circle. A couple of weeks later, Papaito was sent to jail for a mobile phone incident. On their way home from school, he and his girlfriend were walking behind a man whose phone accidentally dropped to the ground. Papaito, who was still without a phone, grabbed the phone and decided to keep it. Meanwhile the owner of the phone got word of who had "found" his phone and filed a complaint in the hopes of recuperating it. Papaito and his girlfriend were immediately arrested and spent three weeks in jail waiting for their court hearing. When the day came, the judge made fun of the couple, saying that they might even end up spending the entire summer in jail for a phone that he would himself not even bother bending down to pick up. It was a Nokia 5110, a phone at the very bottom of the range. The couple was released on the spot after agreeing to pay a fine to the plaintiff.

Suspicion and Envy

"I have no friends. I know a lot of people, but I have no friends," was the reply almost everyone gave me when I inquired into their friendship networks. The young people I worked with claimed to have many *conhecidos* (acquaintances) but only a few *amigos* (friends), people they could trust and people

21. Dino's Bar was a very popular bar on Tofo Beach run by Dino, a Mozambican, and his South African wife. The place was famous for its monthly full moon party, which attracted tourists, expats, local "Beach Boys," and some young people from the city. Drinks were twice as expensive there than in the city but the gang knew how to smuggle in cheap rum.

who, in turn, could also trust them.[22] Relationships based on mutual trust between friends, but also between relatives and intimate partners, were seen as the exception rather than the norm. This is not to say that relationships were not important but rather that they needed to be juggled with caution. The way young people spoke of their networks reminded me of David Webster's (1975) ethnography of kinship and politics in Chopi society. Webster, who conducted research in Inharrime, south of Inhambane, in the 1970s,[23] was struck by the degree of individualism he observed.[24] After sketching the indeterminacy of Chopi residence patterns, which contributed to the fluidity of allegiances, he argued that social survival in Inhambane rested on the development of "network management skills" that, in turn, contributed to high degrees of individualism. Webster wrote in a different era, not only at the political level—Mozambique had just gained independence—but also at the theoretical level, and his transactionalist analysis and the issues it raised are no doubt a little dated. Still, I often caught echoes of his portrait of Chopi society in the narratives of my interlocutors. I repeatedly heard Inhambane youth describe life as an individual battle in which they could rely on no one but themselves. Liberdade residents knew to never let their guard down and to avoid being fooled by the pretense of intimacy. They knew all too well that harm often comes from those closest to them. The one who empties the drunken man's pockets spent the evening drinking with him at the same table; witchcraft accusations usually pit relatives against each other; and love triangles often involve individuals who know each other well. Distrust, then, was the default approach in most social interactions as everyday life was entangled in a "hermeneutic of suspicion" rooted in a violent past (Ferme 2001).

Envy was blamed for all sorts of social afflictions. It was understood to drive petty crime and interpersonal violence, and to keep the traditional healing industry afloat. Proverbs explicit about the ills of envy embroidered on decorative mats hung on the walls of houses and *baracas*. Reflecting on the socioeconomic impacts of envy, Kenneth once suggested we open an NGO aimed at fighting envy in Inhambane! (See also Sanders 2001: 173.) The fact that *inveja* (envy) and *odio* (hatred) are often used interchangeably in Mozambican Portuguese also gives a sense of the intensity of these sentiments.

22. Simmel (1906), writing about acquaintances, wrote: "That persons are 'acquainted' with each other signifies in this sense by no means that they know each other reciprocally; that is, that they have insight into that which is peculiarly personal in the individuality" (452).

23. Webster's research was based in the district of Inharrime, south of the city of Inhambane, which is predominantly Chopi.

24. He stated that that there was "more individual expression and freedom in Chopi society than [was] permitted in most traditional societies in Southern Africa" (Webster 1975: 340).

Envy often manifested itself through *ambição*, or the desire for something lying beyond one's means. *Ambição*, unlike its literal translation in English, "ambition," has a negative connotation in Inhambane where it is viewed as pretentious, often unrealistic, and, more importantly, as inciting people into doing "ugly things" (*coisas feias*), such as reverting to witchcraft or theft.

The recent scholarship on brokerage in Africa has highlighted the creativity of brokers as they straddle "the borderline of legality" as well as their part as "both product and producers of a new kind of society" (James 2011: 335). In southern Mozambique, this "new kind of society" has seen the erosion of traditional paths to material security, changing from a "labor-scarcity economy" to one of "labor surplus" (Ferguson 2013). In the petty crime economy, the appropriation of resources operates not so much through productive or reproductive labor but rather through what Ferguson (2015) has called "redistributive labor." Such a sketch should not, however, occlude the way young people commonly navigate between different spheres of activity, dabbling into different things at different times of their lives, of the year, and of the month. In anthropology, a more dynamic framework is preferred to the more rigid dichotomy between the formal and the informal economy (Hart 2010; Hull and James 2012). Individuals living in social environments marked by material insecurity often participate in various, at times overlapping, at times contradictory, fields that command and complement each other in myriad and often unexpected ways. Concepts such as social navigation (Vigh 2006), tactical agency (Honwana 2005), and zigzag capitalism (Jeffrey and Dyson 2013, building on J. Jones 2010) are particularly useful in making sense of how people address and engage with material insecurity. These concepts clearly highlight the importance of mobility in successful living. The phone is such a powerful tool precisely because it helps compress space and time, but also because it itself circulates *and* helps set things in motion.

Gendered Economic Spheres

In this chapter, I discussed alternative modes of access to money and things, mainly from a male perspective, and the role mobile communication along with mobile phones themselves, as a form of quasi-currency, play in these exchanges. Although some women also partook in criminal activities—for instance, a young woman I worked with was jailed for stealing the credit card of a tourist with whom she had been intimately involved—it nonetheless remained a male-dominated sphere. And although men struggled to live up to mainstream ideals of masculinity that cast males as independent breadwinners, it was striking how resilient this hegemonic model proved to be (cf. Ferguson

2015). Indeed, men have held on to a model of masculinity that emerged in a rather different socioeconomic context when income was available and consumption not so frenetic.

In this context, the onus was increasingly on women to take charge of household subsistence by engaging in urban agriculture and petty trade, or working as domestic servants. Young women were, as it were, contending with a different set of occupational stigma to those of their male peers. Though opportunities within the formal sector remained scarce, they could more easily build on their dependent position to tap into kin and intimate networks. Women retained much control over the domestic sphere, while also redrawing the boundaries designed to relegate them to the household. Few had made it past their twentieth birthday without getting pregnant, and several were second-generation single mothers.

The importance of female-headed households and the reliance of many on female-generated income were but two of the most visible manifestations of the postsocialist, postwar economy. Male authority continued to guide gender relations, justify polygamy and other male privileges, inform the ideal of the patrilineal household, and dictate normative behavior, especially regarding women's sexuality and freedom of movement. In short, it remained part of what Michael Herzfeld (2004) has described as "larger discursive universes in which everyday experience is embedded and from which social actors draw legitimacy" (318).

Women, have, on the whole, proven far more adaptable (Cole 2010; Silberschmidt 2005; although see Johnson-Hanks 2002) and overall more successful at redefining themselves than men who have been far more reluctant to try out new ways of being men. As I discuss in chapters 5 and 6, it is, however, somewhat ironically often by playing on their subordinate status that women have been able to redefine themselves and escape, even if only momentarily, from certain forms of domination. Young men's failing role as breadwinners has encouraged women to go look elsewhere and seek financial support from more secure, generally older, men. Indeed, young women have developed crafty ways to deploy their sexuality to their advantage. And, if sexuality has come to play a more central role in expressions of masculinity (Manuel 2008; see also Silberschmidt 2004), so has femininity become ever more sexualized. Women have long been valued for their productive and reproductive powers, but they are increasingly defined by their "bodily capital" (Wacquant 2004).

The struggles of adulthood in sub-Saharan Africa, like the demands of intimacy more broadly, are imbued with gender-specific inflections. Scholars of Mozambique, in particular, have shown how men and women operate in "separate [economic] spheres" (Pfeiffer, Sherr-Gimbel, and Augusto 2007). Unlike women who could tap into broad intimate networks—a theme developed in chapter 5—the networks on which young men could rely for assistance were also very different in scope and nature, even if there was, as discussed in the final section of chapter 5, some overlap. These differences aside, success in either sphere—the criminal or the sexual—was reliant on one's ability to juggle visibility and invisibility. Because I find it more useful to situate young women's sexual networking practices within an intimate—rather than a sexual—economy, I start by exploring the redefinition of intimacy underway in postsocialist, postwar Mozambique through the lens of mobile phone practices and with a focus on young people's experiences with love and jealousy, before turning to the workings of the intimate economy.

4

Love and Deceit

When I returned to Inhambane after nearly a year of absence, one of my ex-neighbors, a flamboyant old man known as Takdir, called me over. He looked concerned. “They’ve had to send you back?” he said in a way that sounded more like a compassionate observation than a question. “Let me help you; let me tell you the answer to your research,” he told me, before adding with assurance: “Youth here use mobile phones to *namorar* [to develop intimate relationships]. Some say that phones help with business but that’s not true; phones are for *namorar*!” Of course, the youth of Liberdade used their mobile phones to do many things, whether it was to keep in touch with distant relatives, inquire about this or that, or light up their way at night. But, as Takdir pointed out, very often they used their phones to *namorar*: to flirt, persuade, coordinate romantic *rendezvous*, or insult a rival.

In this chapter I examine the negotiation of intimate relationships and the adoption of new ideas about intimacy through the lens of mobile phone practices. As I show below, young people were using their phones as powerful tools of persuasion and cajolery, often in ways that fostered open-endedness. Many also felt that mobile mediation was transforming intimate relationships by facilitating infidelity and breeding intimate conflicts, as well as by encouraging the commodification of intimacy. Leaving debates around the commodification of intimacy for the next chapter, what I propose to focus on in this chapter are the ways in which experiences with love and jealousy are being redefined in the new intimate spaces opened up by mobile communication. After situating mobile phone seduction historically and in relation to broader courtship practices, I turn to the phone as a catalyst for intimate conflicts and show how, through the interception of incriminating phone calls and text messages, the adoption of mobile phones has also opened up new discursive spaces within which partners can negotiate the terms of their relationships. What the chapter offers, more broadly, is a reflection on the relationship between mediation, authenticity, and the materiality of truth.

Flirtation and Pretense

Built along the abandoned railroad track, Mafurera market consists of a long corridor of lined-up stalls which sell a variety of foodstuffs, cosmetics, and secondhand clothing. On one of my early morning visits to the market, I bumped into Marta who, like me, was shopping for breakfast. She told me that she was on her way to buy bread at a small bakery on the other side of the market and invited me to join her. As we passed other bread sellers along the way, Marta tried to convince me that the bread at the more distant bakery was well worth the extra mile, but she sensed my skepticism and added: "I have to give everyone a chance to get a good look at me," at last divulging her parallel agenda. Parading in a cool beach sarong that drew attention to her voluptuous hips, Marta sauntered ever so slowly through the market. We stopped for a moment at a stall selling the latest fashion where I heard her apologize to the young man working there for having castigated him a few days earlier. As we walked on, Marta filled me in on what had happened.

On his way back from a shopping trip in South Africa, the man phoned Marta to tell her that he wanted to see her as soon as he arrived in Inhambane later that evening. Marta agreed to meet with him and "gave him morale" whenever he updated her on his progress by confirming that she was waiting for him. It was getting dark when he told her that he was finally passing Cumbana, a town about sixty kilometers south of Inhambane. Marta told him how much she was looking forward to seeing him but, after hanging up, she switched off her phone and went to bed. "I didn't feel like seeing him anymore," she told me. "But weren't you worried that he would come knocking at your door?" I asked. "Of course not," she replied, "a guy can't just turn up at a girl's house. What would my mother say? We're safe at home!" But why not have the courtesy to let him know that she had gone to sleep, I wondered. Marta explained that she simply felt like stringing the seller along. Although she had no intention of seeing him that evening, she was reluctant to reject his invitation and believed it was preferable to withdraw at the last minute. Offering an important piece of advice, she told me that it was vital to make men wait as long as possible before putting out. To him, however, she had made up a story about period pains, a tactic that women readily admitted to using more or less frequently whenever they wished to remain sexually unavailable. Marta's excuse was plausible and open to interpretation. It was aimed at fostering a productive, potentially enticing, sense of uncertainty, or an open-endedness that precluded a definitive outcome.

The following week, however, the tables had turned. The seller had grown cold and had even stopped responding to Marta's *bips*. Marta eventually gave up and rationalized the affair by insisting that, in any case, she had only been interested in taking advantage of him. As she explained, what had caught her eye was not so much the man himself as his lucrative clothes business. "I never truly liked him" (*nunca gostei dele de verdade*), she told me. Marta, who was in her early twenties at the time, was living with her retired mother on a small plot of land near Mafurera market. The household relied on remittances from relatives working in other parts of the country to supplement the mother's meager pension. Marta had a son with Abibo, a married man who lived in a nearby neighborhood and who also contributed a little every month. In contrast with the language she used to talk about the market seller, she spoke of Abibo in very affectionate terms and told me that she was unable to resist him, however hard she tried. She described her feelings for Abibo as authentic, as true or *verdadeiro*.

When Marta admitted never truly liking the market seller, it was not the first time that I had heard someone talk about flirtation as the expression of desires that were not necessarily genuine. I had often heard both men and women speak of how difficult it was to know whether someone's feelings were "*sentimentos verdadeiros*" (true feelings), and, in turn, emphasize how important it was to proceed with caution, to conceal one's feelings, especially if these were authentic, as transparency would place one in a position of vulnerability. As Jenny explained, "I've discovered that it's best not to let the man you fancy know that you're a big fan of him (*maningue fã do gajo*)." Her friend, Maizinha, added: "When you show [a man] that you like him, that's when it starts going wrong, that's when he starts losing interest." In this social environment marred by profound suspicion and deceit, little, if anything, was ever taken at face value. The consensus was that it was impossible to know with certainty whether someone's sentiments were genuine or simply driven by ulterior motives, whether one loved "for real" (*se ama de verdade*) or whether he or she was only pretending to love (*disfarçar*). As Inocencio, the young man I quoted earlier, explained during a heated conversation about infidelity: "You girls are used to being lied to and you force us to lie. Women prefer to be lied to, and when you speak the truth to them, they don't believe you, so it's better to lie." The demands of intimacy, in other words, called for ambiguity and ambiguation. While men complained that women pursued material interests above all else, women, for their part, felt that men were always trying to lure them into bed under false claims of true love. There was a shared understanding that intimate relationships were inevitably characterized by deceit as feelings were either downplayed, falsified, or all together fabricated so as to gain some sort of benefit, and everyone saw the phone as amplifying these trends by making it easier to persuade, deceive, and accumulate intimate partners.

The uncertainty characteristic of courtship, then, was, as Matthew Carey (2012) notes in his analysis of a flirtatious encounter in Morocco, "not merely something to be worked around, but rather something that people work with" (190). Intimate uncertainty, in other words, also yielded uncertain outcomes while fostering potentially productive open-endedness. In a similar vein, Vincent Crapanzano (2004) usefully describes flirtation as the display of possibility. He writes: "Human play includes the optative, the subjunctive, the negative affirmation, and the affirmative negation" (146). For young adults in Inhambane, these different tenses and formulations were also conjugated according to the broader vagaries of everyday life.

In Pursuit of Romance: From Kin to Mobile Mediation

Mobile phones have made their way into mainstream Mozambican society at a time of important socioeconomic transformations and are also understood to have radically transformed courtship practices. In southern Mozambique, courtship has historically been mediated through, and bound up with, patriarchal authority. Inhambane is part of the patrilineal south where descent is traced through patrilineal filiation, marriage alliances usually involve the exchange of *lobolo* (bridewealth), and residence tends to be patrilocal (Arnfred 2011: 85). In contemporary Inhambane, although we find much variation in terms of living arrangements, including a high proportion of women-headed households, the patrilineal, patrilocal household still holds currency as a valued ideal. Older Inhambane residents recalled a not-so-distant past when marriage was "a serious matter." Their narratives highlighted the importance of kin mediation in intimate affairs and detailed how men would express their interest in a particular woman to their male kin who, provided they approved, would then approach the woman's family and thus set in motion a lengthy process of negotiations.[1] The Swiss missionary and first ethnographer of southern Mozambique, Henri A. Junod, also hinted at the importance of choice in marital preferences and mentioned the recourses available to women who, for whatever reason, had strong reservations against a proposed suitor (Junod [1912] 1966: 106). In line with the classic anthropological literature,

1. The historian Patrick Harries (1994) has described a similar custom in his work set in rural southern Mozambique.

marriage was, however, first and foremost understood as a union between two families and *lobolo*, the transfer of women's productive and reproductive powers.[2] It was also through marriage that generational hierarchies were maintained and eventually challenged when labor migration enabled young men across the region to secure themselves the means to acquire wives (West 2005: 104).

Mark Hunter's (2010) work on love and sexuality in neighboring South Africa adds to classic regional kinship studies by focusing more closely on the affective dimension of marriage.[3] Challenging the idea that romantic love is a modern phenomenon (cf. Giddens 1992) and foreign import, Hunter goes back to nineteenth-century Kwa-Zula Natal to show how love in South Africa has traditionally been expressed through cooperation and mutual assistance (Hunter 2009: 136). If migrant labor allowed men to acquire bridewealth and thus emancipate themselves from the control of their elders, it also, as Hunter argues, "provided men with an economic basis for ideas of choice central to the ideology of romantic love" (ibid.: 141). Hunter's reading of historical sources moves emphasis away from the common image of migrant labor as pitting young men against their elders to migrant labor as driven, at least to some degree, by romantic intent. And while *lobolo* can, indeed, be understood as an important institution that symbolizes the transfer of responsibility of a woman's guardianship from her father to her husband, it also symbolizes women's social value (ibid.: 137). As such, Hunter writes, "over the twentieth century, ideas of love became embroiled with men's emergent role as providers" (136). As mentioned earlier, and as detailed later, love across southern Africa has remained very much predicated on men's ability to provide.

In contemporary Inhambane, *namorar*, which I will translate as "dating" for want of a better word, has to be understood against the backdrop of kin-mediated intimacy. My research participants sometimes insisted that *namorar* was not so much a departure from traditional forms of intimate relationship but rather an extended step, which allowed the couple to get to know each other better before committing to marriage. Sometimes, however, they felt that their ways of understanding intimacy were radically different from those

2. As anthropologists have shown, practices like bridewealth that appalled Europeans could be understood in terms of the transfer of "productive and reproductive rights and responsibilities within and between age groups," but there is a fundamental emotional dimension to these exchanges that has long been overlooked. Women are proud of how much their husband and his family paid for her, not unlike a woman in North America may be proud of her engagement ring (Thomas and Cole 2009: 22).

3. See, for example, Kuper's (1982) *Wives for Cattle* and Krige and Comaroff's (1981) edited volume on marriage in southern Africa.

of previous generations. Crucially, and whether or not it eventually morphed into anything more durable, *namorar* bypassed kin mediation still so central to marriage negotiations, and called for other forms of mediation.

In conversations on changing courtship practices, the phone often arose as an important reference point. The introduction of mobile phones was seen as resolving several technical challenges commonly encountered in the pursuit of romance by making it both more direct and discreet. People in Inhambane remembered how prior to mobile phones, intimate relationships relied on different forms of mediation. Elderly people in Inhambane recalled, for example, how engaged couples would stand on either side of a tree and exchange kind words. Some older residents suggested that in their days, young people would sometimes send each other love letters but that the time delay made it an unpopular alternative. Instead of raising the much more important issue of illiteracy, they assessed letter writing in contrast to real time mobile communication. Others spoke of how young men would "hunt" women by hanging out in the alleys around the market,[4] and suggested that, as a public space of high visibility, the alley offered serious restrictions as a courtship space. Not only did people rely on uncles for marriage negotiations, lovers also depended on the assistance of young children serving as messengers to exchange love letters. If many recalled sending young children to transmit messages and the risk of them spilling the beans, they were grateful for how mobile communication allowed them to avoid such intermediaries. In short, mobile communication was seen as transforming seduction into a less mediated, more personal, much easier—in theory, at least—and, ultimately, more private experience. As a young man simply put it: "Before, when I wanted to talk to a girl I liked, I would risk getting beaten by her brothers or her boyfriend. Now, I just phone her." Such a statement may sound trivial but in a place like Inhambane where most have gone from no phone to mobile phone, where people live in close proximity, mostly outdoors, and where privacy is scarce, mobile communication has quite literally opened up new spaces within which individuals can pursue their interests with some degree of discretion, and with vital implications regarding respectability, the

4. As a growing number of young women pursue secondary education, the time they spend away from the control of family members has also increased. Women therefore have new opportunities to meet men. In fact, there are a number of songs that speak of women being hounded by older men on their way to school. Although it is still early to determine how education will shape Inhambane women's access to formal employment, the effects of going to school every day are more readily observable. Temporarily freed from the surveillance of male kin, young women can now encounter new opportunities "along the way" (cf. Thomas 2006: 182).

accumulation of lovers and suitors, and the overall smooth management of intimate networks. At least until it all backfires (see below).

Looking at the history of media and communication, communication technologies usually have an impact on ways of being and relating, and on courtship practices in particular. Letter writing, for instance, has, as Karin Barber (2007) writes in her study of textuality and personhood, played "a special, central role in the imagination of new modes of privacy and personhood in many cultures" (177). Letter writing gave men and women a medium through which they could negotiate intimacy while experimenting with new ways of presenting the self. In her analysis of paternity trials in East Africa, Lynn Thomas (2006) gives the example of a young pregnant woman who used mail correspondence with a man away at college not only to make claims about paternity, but also, through her epistolary skills, to showcase her modern outlook. Letter writing opened up spaces within which new (and sometimes not-so-new) ways of being and relating could be imagined and tested. In *Becoming a Woman in the Age of Letters*, Dena Goodman (2009) similarly shows how epistolary practices participated in the redefinition of femininity and the expansion of privacy in eighteenth-century France as women used letters to convey their innermost thoughts and feelings. She also focuses on the materiality of letter writing to highlight how writing desks with their locked drawers and secret compartments were themselves enhancing privacy. With new intimate possibilities and opportunities, however, also often comes dissatisfaction. Laura Ahearn's (2001) account of love-letter writing in Nepal nicely shows, for instance, how changes in courtship and marriage practices following the acquisition of literacy skills can also foster new kinds of disappointments.

Like letter writing, mobile communication has encouraged new ways of imagining and pursuing intimacy. In some cases, the phone has been integrated into older forms of marriage negotiations and match-making (Tenhunen 2008); in others, mobile communication has contributed to the random expansion of intimate networks such as when users attempt to establish contact by sending messages out to "wrong numbers" (Andersen 2013; Carey 2012). In many cases, mobile communication has helped subvert particular forms of authority such as parental control (Ito, Okabe, and Matsuda 2005; Lin and Tong 2007) and gender norms (Maroon 2006). My exploration of the phone's impact on intimacy in Inhambane starts by looking at how the phone has been turned into a powerful tool of cajolery.

In a place where men are expected to demonstrate their interest, and in turn their worth, by making known their ability and willingness to provide, the phone can prove particularly useful. By calling and sending regular text messages, for instance, men can convince women that they are genuinely interested. And, in a place where women are, for their part, expected to play a relatively passive role in courtship, mobile communication offers some degree of poetic license to women who may otherwise lack assertiveness in face-to-face encounters. Here is an example.

Mikas, a married man in his early thirties, was determined to seduce Fina, a young woman he had met at church, but Fina kept refusing to give him her phone number. This failed to deter Mikas, who pressed on, so much so that Fina eventually gave in. As Mikas soon found out, however, Fina had given him a false number. Mikas was impressed by her audacity and even more resolute to get hold of her number. In some cases, men would take a woman's phone and send themselves a *bip* or missed call so as to "catch" the number but, in this case, Mikas finally succeeded in convincing one of Fina's friends to share the coveted number. Unable to escape his calls and messages, Fina was eventually charmed by Mikas's determination and decided to play along. At first she kept the exchanges light and formulaic but, a few days on, she started placing requests for airtime and even asked him to buy her a pair of sexy underwear. The explicit nature of Fina's messages was putting Mikas at risk of getting in serious trouble with his wife, were she ever to intercept the exchanges. Mikas, who initially found the conversation titillating, eventually lost interest. He had been taken aback by Fina's unexpected brazenness and also admitted that he was in no financial state to start subsidizing other women's undergarments.

Courtship was commonly described as a measured, contrived, and protracted negotiation that entailed careful reflection from both parts. In fact, men themselves spoke of the women they were hoping to seduce as "projects" in recognition of the perseverance required in courtship. Typically, this is how it works: The man approaches the woman he is interested in and asks her to *namorar* with him. The woman initially expresses a lack—often a total lack—of interest but is nonetheless given some time to consider the offer. Both then investigate each other's behavior (*pesquisar o comportamento*). Women are known to stall the process for as long as possible, thus forcing suitors to further demonstrate the authenticity of their interests through small gifts.

When I asked Inocencio how he went about seducing women, he said: "I am not a romantic, I don't have romantic words. I am straightforward, I'll say, 'I like you' [*gramo de ti*]. So I start by saying, 'Hi, how are you' and then I tell her how I feel about her. I practice at home beforehand, if it's a girl I really like. But if it's a girl I just flirt with because she's a woman and I'm a man, then I just talk to her. Either way, I make sure I get her number."[5] By spending money on phone calls and by being generous with airtime, men were using mobile phone communication to prove their interest and commitment in a tangible, even quantifiable, manner.

Instead of having to wait to be "hunted down" on their way to the market, women, for their part, could play a more active role in courtship. Indeed, many felt that they could more easily express their desires and also place requests for specific things through mobile-mediated exchanges rather than in face-to-face encounters. In the previous example, Fina began by displaying indifference but she eventually adopted a much more active and licentious role in her exchanges with Mikas. But as the previous example also shows, women's subversion of gender expectations was not always welcomed. Men were easily scared off by such advances as they clashed with received notions of femininity and decorum—women are expected to play hard to get—and also because they involved a financial commitment that was often beyond their reach. Everyone agreed, however, that mobile phone communication facilitated—encouraged even—infidelity.

Making Sense of Infidelity

Infidelity in Inhambane was awfully common and somewhat expected of men, and increasingly of women as well. It was, however, extremely contested and, when discovered, would generally spark great fury. Infidelity deserves, first, to be situated in relation to polygamy, even though infidelity is understood as pertaining to a rather different domain embedded, as it were, in secrecy and deception. The region has historically been known for its high incidence of polygyny, especially in rural areas dependent on agriculture.[6] Frelimo's attempts to ban polygamy in the 1980s were, in fact, met with resistance and hostility.[7] And, given the link between wealth and access to women, the prohibition of polygamy intruded in the life of prominent individuals who were unwilling to give up the privilege. Frelimo even eventually reversed its stance against polygamy when it undertook a series of reforms in the late 1980s.

5. As mentioned in chapter 2, young men might also use the expression *latar uma dama* (literally, put a young woman in a can or bottle [with conversation]), to seduce them.

6. According to research conducted in the area south of Inhambane in the 1970s, 36 percent of male informants reported having lived in a polygamous union at one point in their life (Webster 1975: 150).

7. Arnfred (2001) has shown that although the Party's stance against polygamy was meant to protect women, it had the reverse effect as it prevented second wives from getting formal recognition.

Polygamous households were, however, often fraught with conflict, especially in cities where co-wife rivalry was compounded by the particular challenges of getting by in an uncertain urban environment and by the appeal of competing notions of intimacy. In an attempt to mitigate household conflicts, men in contemporary Inhambane often kept their different wives in separate compounds, even if doing so, in turn, enhanced the economic (cf. Arnfred 2001; Locoh 1994) and affective pressures they faced as polygamous men. Here is an example: Janu, the youngest of six children, was living with his first wife and their children in his parents' compound. A few years ago, when his lover fell pregnant, Janu and his parents agreed that the sensible thing to do was to take her in and to build her a small house next to his first wife's house. The lover thus became his de facto second wife. From the beginning, the two women were unable and unwilling to get along. The first wife, who already had three daughters with Janu, worried she might lose her husband's affection were the other wife to give him a son. As their sister-in-law put it, both women were engaged in "procreative competition" with each other and the constant bickering carried on for a couple of years. The sudden death of the second wife's toddler then raised suspicions about the first wife's malicious intentions. Something had to be done. Janu decided to rent out another property some distance away where he relocated his second wife and their new-born child. Janu earned a modest salary as a barman in one of the city's nightclubs and although he was far from comfortable, by any standards, unlike most young men in the neighborhood who were unemployed, he at least had a reliable source of income. I knew of only a small number of young men who were able or willing to take on the responsibility of having a wife, let alone two. Some even felt that having a girlfriend was beyond their reach.

Mozambicans have long borrowed from various repertoires to imagine and construct intimate relationships, and even a cursory look at popular culture shows how love and jealousy are favorite topics of discussion (cf. Thomas and Cole 2009). According to Linda van de Kamp's (2012) account of Pentecostal love therapies in Maputo, intimate relationships were placed on a love continuum. The range spanned from *namoro* relationships between boyfriends and girlfriends, which were meant to rest on love and respect, to the *saca-cena* (one night stands) that were inspired by lust rather than true sentiment, passing by the *pito/pita* relationship which was usually more lasting but nonetheless informal.[8] Young people in Inhambane made similar distinctions

8. Much has been written on these different types of relationships in Mozambique alone and on the way Mozambican youth have been reimagining intimate relationships and sexuality (Groes-Green 2010; Manuel 2008; van de Kamp 2012).

that they defined in almost identical terms, but in practice a slightly different portrait seemed to emerge. Although young people commonly said that they hoped to one day marry someone they loved, their choices of boyfriends and girlfriends (as potential husbands and wives) were still very much guided by deeply rooted notions of respectability. A good potential wife was generally a tightly controlled young woman from a reputable family who would make an upright wife and responsible mother. Lynn Thomas and Jennifer Cole (2009) similarly note that "although emotional and physical attraction commonly animate courtship and affairs, it is not valued as a solid foundation for marriage" (7). The more formal relationships between *namorados* as well as between husband and wife were also commonly ridden with conflict over finances and jealousy in ways that often precluded intimacy. Both men and women found that they often had to look elsewhere, *fora* (literally "outside"), to satisfy their sexual and affective needs and desires. When discussing what encouraged them to develop "outside" intimate relationships, young men spoke of their desire to spend time with a woman with whom they could *desabafar* (see chapter 3); that is, let down their guard for a moment and relieve their stress. In fact, the intimate partners that were referred to with the most affective terms were usually lovers (*pito/pita* or *damo/dama*). The person one longed to spend time with was often the person one could not be seen with because they were officially involved with someone else or because the relationship was, for whatever reason, socially unviable. In postwar, postsocialist Mozambique, not only have new forms of intimacy become imaginable, so have economic conditions, as write Thomas and Cole (2009: 27), "made it difficult for one relationship to fulfill all emotional and material needs."

Women in Inhambane were all too aware that married life could be rather dismal. Tellingly, of all the young adults I knew, the only ones who described themselves as unhappy were married women.[9] Anaty, a young woman who had recently left her *sograria* (the house of her in-laws) after living there for just over a year following the birth of her son, complained about how difficult living with her in-laws had proven. "I was ugly and looked old when I was living there as my mother-in-law kept me so busy all the time. I never even had time to braid my hair. I had to keep my nails short. Meanwhile my husband would spend all his money on other girls!" Anaty was secretly relieved when her husband grew tired of her and sent her back to her uncle's house where she had previously been living.

9. "Married" here, as earlier, means seriously committed rather than officially married.

Extracts of a song sung by a mother to her daughter before her wedding, more than a generation ago, nicely highlight the tribulations of married life, echoing Anaty's laments:

Go away my daughter go away
You do not know to where you are going
Weep my daughter weep
You will suffer in the house of your husband
You will have to work a lot
May not even sit down when you are eating
You will be everybody's servant
Fetching the water
Chopping the firewood
Washing the clothes for everybody else
You will have to heat the water
Preparing the bath for your husband
For his father, for his mother, for his aunt
The water is not enough
You must go to the river for more
Working, working
Your husband will beat you
Your mother in law will call you a thief and a liar
You who never stole nor lied
Weep my daughter weep
You will suffer in the house of your husband

When this song was recorded, women's options were rather limited.[10] As mentioned earlier, if they felt very strongly about a proposed suitor, they could influence the marriage negotiations (Junod [1912] 1966: 106), but refusing to get married was not a conceivable alternative. Today, married life is further tainted by the way men devote ever more time, affection, and resources to lovers.[11] At the same time, however, as men struggle to pay *lobolo* and, more broadly, to formalize these intimate relationships, women, for their part, have managed to remain somewhat freer (cf. Hunter 2009: 145). Telling

10. The song was recorded by Signe Arnfred in Maputo province in 1982 (Arnfred 2001: 12). See also Marshall (1993: 67) for a variation on the same theme and Lizha James's "Nita Mukuma Kwini" pop hit for a modern version, available on YouTube, https://www.youtube.com/watch?v=k_mOUDfCF68, accessed July 19, 2008.

11. One married woman told me about how her husband's lover had once referred to her as their "domestic servant" (*empregada*). The "rival" had added: "While you're at home cooking and doing laundry, your husband's out with me having a good time!"

of these transformations, there has been a steady decrease in the number of marriages recorded in Inhambane, alongside a notable rise in the average age at first marriage.[12]

From Dakar to Nairobi, we find young people who define themselves in contrast to their parents' generation through their intimate practices (van Dijk 2012: 153; Spronk 2009). As Rijk van Dijk (2012) notes, however, inequalities in terms of access to foreign models of love and intimacy, and to the resources needed to make these ideals even remotely imaginable, should not be overlooked (142). Nor should the sociohistorically specific ways in which these models are locally appropriated. As young people embrace ideas of romantic love, gender equality, and exclusivity—all seen as markers of modernity (cf. Thomas and Cole 2009: 5)—the adoption of mobile phones plays an ambiguous role in the realization and performance of such ideals as it makes it easier to persuade, deceive, and accumulate partners. As entrenched conceptions of femininity, masculinity, and intimacy are sieved through novel registers of romantic love, jealousy remains the main driver of intimate conflicts.

Jealousy was commonly rationalized as an unequivocal proof of love. For instance, although Rafael was confident that his girlfriend was not cheating on him, he still continuously questioned her fidelity, "for her to know that I care . . . also to create conversation," he explained. In some cases, however, jealous sentiments could translate into physical violence. Angela once told me, after showing me her bruised thighs, "If my boyfriend didn't love me, he wouldn't be bothered when he sees me talking to other men, he wouldn't bother beating me if he didn't care about me." Speaking from the perspective of women involved with married men and adding a caveat to the link between love and jealousy, Maizinha, another young woman from the neighborhood, explained:

> One who isn't jealous doesn't love but then there are some who are jealous just to create conflict. If a woman makes problems with me because I'm with her man, I'll stay with him but if she tells me nicely, I'll leave him alone. Ya, but if the woman slaps me, eeh, I'll really go out of my way to please her husband!

Men were, on the whole, reluctant to give up their privilege to accumulate intimate partners. As Hunter (2002) points out, despite the important differences

12. Data I compiled at the Civil Registry in Inhambane suggests that just after independence in 1975, the average age of first marriage was 27.23 for men and 21.70 for women. Since then, it has risen progressively to reach 34.4 for men and 28.7 for women in 2006. Between 2000 and 2006, there were on average 60.1 civil marriages performed, a sharp drop from the 81.5 average marriages per year of the 1990s, and the 91.6 of the 1980s, despite the latter being the war years.

that exist between polygamy and more contemporary multi-partner relationships, men commonly conflated the two when justifying their prerogative to have multiple partners (116). In a popular song by the Mozambican artist Ziqo entitled "*Casa dois*" (literally "second house"), the husband warns his wife that if she gets upset with him, he'll simply go to *casa dois*. Indeed, men commonly used the double standard that they enjoyed as a means to control women. "Women are plentiful" (*mulheres não acabam*) was a popular saying that men sometimes brought up, somewhat facetiously, to suggest that by having many partners, they were in fact acting out of compassion (see also Arnfred 2001: 48). There was some truth to the saying. According to the provincial population survey, there were twice as many women as men aged between twenty-four and twenty-nine. This, however, mainly reflected work migrancy (up until the fourteen-to-nineteen cohort, the male/female ratio is roughly 1:1), an essentially rural phenomenon.[13] In the city of Inhambane, however, men only managed to have many partners of the same age group because at least some of the women also had multiple partners.[14] If men were buying into the myth of demographic imbalance, it was also to prevent them from having to face this harrowing reality, as they valued having exclusive sexual access to women.

Although patriarchal authority has been seriously undermined by state retrenchment and the wider contraction of the labor market, it nonetheless remains the main normative repertoire on which people draw to justify certain practices and condemn others, such that everyone is encouraged to cover up, out of respect, any transgressions. As they negotiate new and not-so-new forms of intimacy, young people in Inhambane are also having to mitigate and dissipate new kinds of intimate conflicts—conflicts that speak of, but also sustain, profound material and relational uncertainty (cf. Taussig 1987). As I show in the following section, the phone often plays a catalyzing role in these dynamics.

13. Statistics of the sort were taught at school without appropriate contextualization, and were later used by the students to validate and excuse various behaviors. Another example was the life expectancy statistic. Students were taught that Mozambican life expectancy hovered around forty. They were not told, however, that this was mainly due to extremely high infant mortality. I heard some young men justify risky behavior, especially unsafe sexual practices, by stating that with such a low life expectancy, there was little motivation to act carefully.

14. Polygyny rests on the existence of excess women and on age differential at marriage (Bledsoe and Pison 1994: 158; Spencer 1998). The young men I am interested in here have relationships with a variable number of young women that are commonly only a few years younger than them and who themselves are also often involved with much older men.

"Breaking Up Because of the Phone": Dealing with Digital Traces of Deceit[15]

Antonio was studying with Ana, his girlfriend, when he received a text message. "It was from a friend," he explained to me a few days later, "but it was a little provocative." Still, he agreed to show the message to his girlfriend after she heard the notification and demanded to see who was contacting him. As soon as she saw the message, Ana started shouting and accusing Antonio of cheating on her. She then grabbed a kitchen knife and stabbed herself.[16]

Most people in Inhambane have, like Antonio, a story about themselves or a couple they know who argued or split up "because of the phone" (*por causa do telefone*), although not all are as dramatic as this one. In fact, such conflicts—some like Antonio, would call them misunderstandings—were described as the phone's biggest drawback. The phone facilitates access to information and not necessarily the kind of "useful information" referred to by endorsers of the ICT for Development perspective (cf. Slater and Kwami 2005; Donner 2008). At times the phone facilitates the circulation of sensitive information meant to remain secret.

Although few, if any, would have done without their phone, many, especially young men, expressed ambivalence toward the new technology. Listening to the narratives of young people, it became apparent that the technology came with hidden costs—costs beyond phone bills. Many had in fact experienced firsthand some of the downsides associated with the technology and, for these young adults, the price of mobile communication was understood as exceeding economic costs. The most burning complaint that they expressed concerned the role mobile phones were seen to play in intimate affairs, namely the conflicts that emerged following the interception of an incriminating phone call or text message like the one Antonio received. These are the conflicts focused on here.

In Liberdade, rumors take little time to spread. Without proof of deceit, however, rumors can often be denied as pure fabrication. As one young man explained, "without proof, I'll still love my girlfriend because people can just be saying that she is seeing someone else to break us up." There is, in fact, profound

15. An earlier version of this section was published in *New Media and Society* (Archambault 2011).

16. Antonio managed to rush her to the hospital where she soon recovered from what turned out to be a superficial wound. The couple, on the other hand, did not survive.

suspicion that others are determined to sabotage relationships either by resorting to witchcraft, by seducing other people's partners, or simply by spreading rumors. The dismissal of hearsay should, however, not be read as indicative of trust between partners; in many cases it rather reflects local regimes of truth. As mentioned earlier, committed individuals are expected to find lovers "far away from home" (*longe de casa*) and to do everything not to be discovered. A good partner is therefore not necessarily a faithful one, but a discreet one, and so long as individuals can feign ignorance of their partner's other relationships, this is what they usually do.

Since the arrival of mobile phones, traditional modes of circulation of sensitive information have undergone important transformations. Indeed, mobile communication has altered the dissemination of information in a variety of ways, some more contentious than others. To start, the technological mediation of interactions has facilitated attempts at falsification and misrepresentation (cf. Burrell 2008). Indeed, the fact that it is easy to lie over the phone was not lost on anyone. "The phone, how it lies!" I often heard people say before adding something along the lines of, "your girlfriend can answer 'yes I'm at home' when in reality she is out and about!" Some attempted to use the phone as a "digital leash" (Ling 2008: 14) and to keep tabs on their partner. As one woman put it: "If you keep phoning your husband, asking how he is and where he is, it's more difficult for him to get up to no good." Some had even designed shrewd strategies to turn the phone into a more efficient tool of control. Samuel, a young man who worked for a demining company and was often away from home weeks at a time, was extremely concerned about his wife's activities during his absence. He would therefore phone her randomly at night and, to confirm that she was truly at home, would ask her to wake up their son and get him to say a few words on the phone.

As a "repository of personal information" (Ling 2008: 97), a phone contains, in call logs, inboxes, sent messages, and even saved contacts, traces of interactions that could be used as material, undisputable proof of infidelity. A romantic text message offered more compelling evidence of deceit than the hearsay of neighbors who might have personal interests in spreading rumors (cf. Paine 1967). Such digital evidence could therefore not be as easily dismissed as rumors. And, as shown below, disconnected phones can also suggest deceit in compelling ways.

"So many couples break up because of the phone," I was told again and again, and Antonio's story, mentioned at the start of this section, was certainly not an isolated case. Tellingly, my survey revealed that 47 percent of female phone owners and 32 percent of male owners reported having fought with their partners "because of the phone." As Inocencio once put it to me:

> I think that the phone sometimes it's uplifting [*anima*] but sometimes it's not uplifting [*não anima*], because of these things of intimate relationships [*namorro*], ya because even if you have a very expensive telephone, you can end up thinking that you wished your phone could disappear. Because it's like this, you meet a girl, and some confuse friendship with something else, and they phone you late at night. . . . Then you, you're with your girlfriend and your phone rings . . . you have to answer and then your girlfriend forces you to answer on loud speaker. "OK, that's fine," you say but then the other girl, *epa*, she starts speaking beautiful words [*palavras bonitas*]. *Epa*, the conflicts because of phones!!

Nor were men the only ones to get caught this way. Johaninha, a young woman who was sitting with us, added:

> Sometimes you receive a very nice message that doesn't come from your boyfriend, but still you start liking the message and then your boyfriend grabs your phone and starts beating you: "How come you have this message here, who is this message from?" You see, when you were just liking the message!

In response to such commentaries about arguing or breaking up "because of the phone," I sometimes ventured that it was not so much the phone as the infidelity the phone uncovered that was the real problem. For my interlocutors, however, this was nonsense. They insisted instead that it was literally the phone that was to blame.[17] The intercepted digital evidence not only meant that it was difficult for the "deceiver" to deny the existence of outside relationships, it also made it difficult for the "deceived" to pretend not to know that their partner was intimately involved with other people. In response to what was experienced as a technological problem, phone users had devised technical solutions such as the delete-all strategy. To prevent problems, most people regularly emptied their inboxes. Joaninha argued that this way her boyfriend might be suspicious—why would someone with nothing to hide delete all messages?—but that at least he would have no concrete evidence on which to buttress his suspicion. She assured me that it was respectful to do so. Some women even preferred not to own a phone so as to circumvent potential conflicts. As Isabella explained, "It's so easy to dial a wrong number, you just have to get one number wrong. But, [my boyfriend] would never believe me if a man phoned and it was a wrong number." Liberdade residents saw themselves as civilized—a recurrent theme in identity discourses—and described phone

17. The offended were sometimes compelled to destroy the medium, either the handset or SIM card, after catching a partner red handed. In his article on communication via intercoms in Indonesia, Barker (2008) discusses the destruction of intercoms in similar circumstances.

investigation as something someone "without education" would do. In fact, many claimed to obey a strict "no touching the other's phone" rule with their partners, a rule that rested on respect rather than on trust. Whether the rule was respected was obviously hard to ascertain—I have given several examples that clearly suggest it was often not abided by—and most therefore regularly cleared their call logs and other folders as a precautionary measure. Many also reported taking their phone with them to the toilet and safely tucking it under their pillow at night. One man was proud to tell me that he had held on to his wife's phone for an entire week during which she had not received a single message or phone call.[18] His friends, however, made fun of his naivety: "Because you don't think that she warned the other guys before handing her phone over to you!" one said, to the approval of the others present.

When people in Inhambane critically described themselves as phone "beginners," as they often did, they meant that they had yet to master the art of respectful phone use, which would have entailed both resisting the temptation of investigating their partner's phone and deleting any trace of suspicious activity on their own devices. If only they had a little more experience with mobile phones, they would not get embroiled in intimate conflicts. Going through a partner's phone constituted a double blow to the politics of respect as not only was it considered disrespectful to do so, it was all the more offensive if it stymied the other person's attempt to negotiate multiple relationships with tact, with respect. Either way, were a woman to confront her partner after finding a suspicious message this way, the argument would most likely turn against her for having taken the liberty to investigate his phone. Unlike evidence acquired through investigation, evidence of deceit that was stumbled upon had added weight as it positioned the deceived in the role of victim rather than investigator purposefully looking for "trouble" (*confusão*).

Intimate conflicts of the sort spoke of deeply rooted expectations of female sexual exclusivity that coincided in form, though not in principle, with romantic ideas of exclusivity and "modem" love. When Benedita asked her husband, Oscar, if she could get a mobile phone, he categorically refused. Determined to own her own phone, Benedita presented the case to Oscar's godmother, a trusted adviser to the family, who soon summoned the couple to her house in order to discuss the matter.[19] At the meeting, Oscar explained that he had once bought a phone for Benedita but that she had lost it out of

18. Ellwood-Clayton (2006) discusses cases in which couples undergo similar tests in the Philippines (363).

19. I was invited along to serve as a witness.

carelessness and that he was therefore not going to buy her a new one. The godmother thought there was more to the story. She was right. After a few beers, Oscar eventually admitted that he had broken the phone in a fit of rage, after finding out that his wife had used it to communicate with her lover, something Benedita did not deny. Oscar, who was crying by then, explained that he suspected his wife still used public phones to contact this other man but that he could not be certain. "At least she doesn't do it using the phone and airtime I buy for her," he added. Benedita justified her behavior by pointing to her husband's own unfaithfulness. "Oscar often goes for days without coming home," she explained. "Once I was hospitalized with malaria and he didn't even know because he was sleeping somewhere else!" It took much convincing for Oscar to admit that his wife had cheated on him and that they were nevertheless still together. Although Benedita was culpable of adultery, she used the conflict to promote her own interests, namely acquiring a new phone and increasing her husband's presence at home.

The interception of incriminating messages and calls was but one of the ways in which mobile phones triggered confrontation. Failing to answer a call or message was sometimes seen as equally suspicious. "When you try to reach your boyfriend and all you hear is '*liga mais tarde*' ['call later,' the operator's message when one's phone is switched off], the first thing that comes to mind is that the person is with someone else," explained Balsa. Reasons for suspicion that would have never existed in the past have emerged with the introduction of mobile phones.[20]

Maria's "phone story" offers a good example of these dynamics. Maria was with her lover when her boyfriend, who works in Maputo and only comes to visit her in Inhambane once or twice a year, phoned her only to hear "*liga mais tarde*." When Maria switched her phone back on a couple of hours later, she received the call of a fuming husband who wanted to know why her phone had been disconnected. "Why not tell him that you were sleeping at the time or that your phone was without charge?" Maria's cousin suggested. The women debated these options and whether they sounded convincing. Meanwhile her boyfriend phoned and they argued some more.

> On one occasion, I was myself the object of suspicion. I woke up one morning to find several messages and missed calls on my phone. The first message came

20. In follow-up research trips to Inhambane, young people were also arguing over the "last seen" tag on WhatsApp, which shows when the account owner was last online. What the person was doing on WhatsApp in the middle of the night was open to interpretation. They could have been chatting with someone or checking when their partner was last online.

from a young man I was meant to meet for an interview the previous day but who had never turned up. Sent at 4:00 a.m., the message asked whether I was asleep. *<ja xtax a dormir?>*

Half an hour later, according to the message log, I received another message from the same number but from Fina, the man's wife, demanding to know why her husband was sending me messages in the middle of the night. *<Uk se passe ke meu marido liga para te de noite?fina>* (What's going on that my husband is phoning you at night?fina)

And a few minutes later, in capital letters: *<PODE NAO ATENDER AGORA MAX AMANHA VAIX ME OVIR BEM MEXMO EU QUERO SABER O PORK DE TROCAREX MESAGEM COM MEU MARIDO ALTA NOITE>* (You may not be responding now but tomorrow you will hear me I want to know the reason why you're exchanging messages with my husband late at night)

Then a few minutes later, but from another number, she sent this: *<Puta come voce troca msgem com meu marido alta noite aky e fina mulher d bino.e voce e uma puta ke so vao te usar pork e a mim ke ele gosta pergute lhe se nao tem xposa de nome fina ta putinha>* (Slut how come you exchange message with my husband late at night here is Fina wife of Bino. You are a slut who will only end up being used because it's me he loves just ask him if he has a wife called Fina)

In some cases, the notion of breaking up or simply "arguing because of the phone" (*brigar por causa do telefone*) was part of a "euphemistic discourse" (Horst and Miller 2006: 169) men and women used to emerge as victims of false accusations. Antonio's message was sent by a friend and the man who might have called Isabella, if only she had a phone, would have been a "wrong number." In other cases, however, and as detailed earlier, talk about breaking up *because of the phone* literally blamed the phone, and its owner's lack of discretion. Within relationships that did not rest on exclusivity or at least the pretense of exclusivity, finding a suspicious message was not necessarily a problem, although both partners may still take offense if other relationships were discovered. It was, however, a different story when it came to more serious relationships in which the woman was expected to be faithful and the man respectful, like *namorados* or married couples. In these cases, the offended had the right to be confrontational.

Popular music offered a constant reminder of the impacts that mobile phones were understood to have on intimate relationships, and several songs that were popular at the time of my fieldwork pick up on anxieties around disclosure. Some songs portray men as victims of false accusations. For example, a very popular song from the Angolan signer Maya Cool[21] entitled "*Prob-*

21. Available on YouTube, https://www.youtube.com/watch?v=he8iU6uPF-Y&list=RDhe8iU6uPF-Y#t=15, accessed January 10, 2016.

lema" offers a vivid example of an intimate conflict triggered by the interception of a compromising text message. It recounts the story of a woman who answers her husband's phone only to find a message saying: "*Querido quero te ver*" (Darling, I want to see you). In the rest of the song, the husband attempts to justify the SMS to his wife by saying that all he did was give the sender of the message a ride. The singer then criticizes young girls determined to destroy families and mobile phones in general. "The phone, these days, is the cause of trouble" (*telemóvel, hoje em dia, é motivo de problema*). The song ends with a dialogue between the man (M) and the young girl (G) who sent the SMS:

M: *Não me liga mais* (Don't phone me anymore)
G: *Vou ligar* (I will phone)
M: *Tenho mulher* (I have a wife)
G: *Não quero saber* (I don't care)

In this song, the man is portrayed as the victim of cheeky young girls, of *meninas atrevidas* (literally, "bold, daring," and more specifically, women who unduly take). One thing the story does not mention, however, is why the man gave out his phone number in the first place.[22]

If some songs such as the one quoted above depict men as victims of false accusations, others such as the following two have men admitting their guilt and engaging in dialogue with the offended partner. A song by the Mozambican artist Doppaz entitled "*Eu Sou Culpado*" (I am guilty) relates the story of a woman who answers her husband's phone only to hear the voice of a woman claiming to be the husband's girlfriend. The furious wife then confronts her husband who denies everything. The latter soon realizes, however, that the proof is too compelling for him to dismiss and so he admits his guilt and asks for forgiveness. The woman thus gains the right and the means to demand an explanation from her husband, who is forced to justify his wrongdoings. Were it not for that phone call, the husband may have been able to brush aside any suspicion as unfounded.

Another song by the Angolan artist Anselmo Ralph, entitled "*Pós Casamento*" (Post-marriage) tells the story of a young man reflectively recounting an argument he had with his wife. The man comes home late one night and finds his angry wife still awake, waiting for him. "Where have you been," she asks, "and how come your phone was disconnected?" The man explains that he was out with his buddies and that his phone battery died. But his wife has already phoned his friends, all of whom denied being with him. "No, no you

22. See also Batson-Savage's (2007) analysis of Jamaican phone-related songs.

are not the man I married, the man with whom I swore love forever," goes the chorus. The man tries other lies but then his phone starts ringing. Already guessing who is calling at 3 a.m., the wife grabs it and hears: "Hello love, are you home yet?" The dispute that follows can only be heard faintly in the background. The song ends with the man asking himself whether it was worth losing his family for an affair.

By providing "proof" of deceit, mobile phones have forced couples to talk (and shout) about issues that may have otherwise been quickly dismissed. Many in Liberdade now have phones with a camera, and this may very well further transform the circulation of sensitive information, especially as a growing number are also joining Facebook.[23] Of course, couples did not wait for the invention of mobile phones to start arguing, but the phone is like what forensic science is to crime scene investigation: it helps buttress accusations. If the phone is seen with ambivalence, it is precisely because of the ways in which it can simultaneously reveal and conceal. Mobile communication has opened up new spaces—virtual and discursive spaces—within which couples can negotiate the terms of their relationships and debate intimacy and gender relations more generally. It is with their phones that young people in Inhambane imagine, debate, and try out new ways of being and relating. These new possibilities are at times a little disquieting, at times exhilarating.

23. Kenneth once showed me a photo of my husband's car in front of the house of a French couple and was convinced that my husband was having an affair with the woman! To my knowledge, he was only helping her weld something. Who knows?

5

Sex and Money

Manuel had recently broken up with his girlfriend and, as he recalled the events leading up to their separation, delved into mobile phone etiquette to make a more general statement on intimate relationships. "Nowadays, relationships are more commercialized," he explained. "If you don't phone back when a girl sends you a *bip*, she'll run to another guy." Manuel was speaking for many young men who, like him, have come to see the phone not only as a catalyst in intimate conflicts, but also as encouraging the reconfiguration of gender relations and of the intimate economy. If these men were faulting women for being ever more exacting, their qualms also spoke of their limited ability to fully participate in this intimate economy. The most critical in this respect were, of course, those who struggled to live up to mainstream ideals of masculinity; those who, excluded from the labor economy, wrestled to fulfill their role as provider and who were not always in a position to phone back when a women sent them a *bip*.

While the previous chapter looked at mobile-mediated courtship and conflicts, and explored affective experiences with love and jealousy, this chapter focuses on the economics of intimacy. Both dimensions—the affective and the economic—are difficult to disentangle, but I find it useful to look at them in turn as they represent particular intimate terrains—infidelity and commodification—that foster their own kinds of uncertainties. Here, I start by showing how phone etiquette acts as a new register used to articulate the reconfiguration of gender relations along with the reworking of ideas of masculinity and femininity underway in Mozambique at the moment.

Whereas living up to mainstream ideals of masculinity has historically been a process that often took men to distant places for extended periods of time, where they accumulated not only capital but also knowledge and mind-opening experience, it has increasingly become performative (Cornwall 2003: 244). Men's inability to fulfill their role as providers has been reported throughout the continent,[1] and it is by looking at mobile phone–induced anxieties that we can start understanding the contours of these young men's struggles.

In her work in India, Ritty Lukose (2012) shows how, through the adoption of various markers of globalization, from hip trainers to motorbikes, young non-elite men mediate globalization and the desire to shine (*chetu*), to be cool. Like in Abidjan (Newell 2012) or in Arusha (Weiss 2002) where young men are also involved in intricate displays of resources, "shining" can backfire when the cracks and fissures it conceals are inadvertently unearthed. If the phone acts as a basic requirement for membership to the world of those who live, the phone can also ironically reveal that one is actually only merely surviving. In other words, while mobile phones play an important part in the way young men and women present themselves as civilized individuals in tune with the trappings of fulfilling lives, the phone often exposes how young men's performance is precisely that—mere artifice—when, for example, they prove unable to answer *bips*, let alone more substantial requests that are often placed over the phone.

In chapter 1, I introduced the various techniques that Mozambicans have developed to communicate free of charge, including some particularly ingenious innovations such as the use of *bips*. These techniques offer interesting examples of how users appropriate technologies in unanticipated ways. As Miller and Horst (2012) point out in their introduction to *Digital Anthropology*, it does not take long for societies to espouse a particular normativity in response to the introduction of new technologies. While appropriation is usually a creative process, what constitutes proper and improper usage—the contexts in which phones should be put on silent mode, if any; whether breaking up in a text message is acceptable or not (Gershon 2010); whether inspecting one's partner's phone is encouraged or seen as an infringement of privacy—soon becomes shared knowledge, even if, like any other social norms, etiquette is also contested through practice. In this regard, *biping* etiquette, as debated below, offers a particularly vivid example of the negotiation of normativity, and of the ways in which norms participate in challenging and reproducing the configurations of power that underpin them.

1. Agadjanian 2005; Cornwall 2002; Honwana 2012; Masquelier 2005; Silberschmidt 2004; Vigh 2006.

The Rules of the Game

Unlike in the North Atlantic world where the adoption of mobile phones and other new technologies has led to a certain polarization of generations (Sciriha 2006; Ellwood-Clayton 2003), the main fault line in terms of usage patterns in Mozambique was definitely gender. Gender differences were particularly salient concerning the allocation of the costs of telecommunication. While they readily recognized the value of mobile communication, young men commonly also reflected on the sacrifices it entailed. Topping up usually involved diverting limited resources away from more collective forms of consumption such as buying food for the family or drinks for buddies. As one young man put it to me, "We feed mobile phones but they're not even persons!" The parallel between mobile phones and dependents is particularly apposite given young men's highly debated inability to provide for women and children. Unlike men who could usually come up with a rough estimate, women commonly found it difficult to estimate how much they spent on top-ups. Such an imbalance stemmed from how the costs of telecommunication were unequally distributed along gender lines. If women did not keep track of how much they spent, it was mainly because they could get other people—relatives, lovers, suitors—to subsidize their phone bills.[2]

These gender differences were a reflection of entrenched ideals of masculinity and femininity discussed earlier that cast men as providers and women as dependents. When *bips* were sent as a request to be called back, to reverse the charges, specific rules applied that replicated this gender hierarchy so that, when communicating with women, if costs were incurred, men were usually expected to pick up the tab. Most of the young women I knew in Liberdade would send *bips* to lovers and suitors on a regular basis. So popular was the practice that it was not uncommon for some to have exceeded their daily allowance of ten free *liga-me* messages by lunchtime. Some of my research participants would sometimes use my phone to send *bips* and, more often than not, the *bip* recipient would call back, despite the fact that the *bip* had been sent from an unknown number. Men, for their part, explained that they were usually compelled to call back whenever they received a *bip*. Because they did not always have the airtime to do so, however, many of them described the experience as particularly stressful. Some explained that this was all the more nerve-racking when the *bip* came from an unknown number, as it kept them

2. I have yet to come across a woman who had given up her phone for economic reasons, whereas it is not uncommon for men to "take a break" from the phone when going through particularly trying times.

thinking, and, given the gendered use of this kind of message, often dreaming about the lost opportunity to answer the request of a young woman.

The gendered rules that encouraged men to cover the charges when communicating with women were often broken depending on the specificities of the situation. Phone etiquette should in fact be understood as a fluid guideline rather than as a rigid set of rules to which one absolutely had to abide by. As the following shows, however, phone etiquette was used to gauge actual practices and those who strayed from the norm risked being ridiculed, perhaps even dumped, as Manuel believed he was. Extracts of an extended conversation I recorded in 2007 will show how a seemingly simple topic of discussion—the etiquette of *biping*—was used to debate broader ideas about gender relations and the performance of gendered identities.

Men versus Women in the *Biping* Debate[3]

On a rainy Sunday afternoon when there was not much to do in Liberdade, I invited over to my house four young people I knew well to *bater papo*. Gina lived nearby with her husband and their son. She had dropped out of school after getting pregnant a few years earlier and, at the time, was working at one of the *baracas* in the neighborhood. João worked as a day laborer. He had never been interested in going to school and was willing to engage in menial labor so as to support his wife and their two children. He had sold his phone a couple of weeks earlier when he had needed money to address a medical emergency. Antonio was a self-proclaimed womanizer, and Tereza was the girlfriend of a wealthy older man well known to all in the neighborhood. Neither Tereza nor Antonio had children at the time. Tereza was completing secondary school, whereas Antonio had recently graduated and was, as he himself put it, "not doing anything."

Tereza was the last one to arrive. She was still on the phone when she came in, seemingly concluding a conversation about a job opportunity. "Something extraordinary!" she told a curious audience after hanging up. From there, the conversation turned to a discussion of gendered experiences with mobile phones. Using Tereza's phone call as a case in point, even though no one present knew the actual details of this "extraordinary" opportunity, Antonio argued that, when weighing the costs and benefits of mobile phones, "for women there [were] always more gains." Addressing Tereza specifically, he added: "You see, you probably didn't even pay for that phone call and it sounds like

3. The following debate was originally published in an edited volume (Archambault 2012a).

you're going to get a job [out of it]!" He further developed his views by bringing in his own girlfriend's phone manners:

> My girlfriend gets money on a daily basis [from her parents], but I only get money maybe once a month and I spend it straight away. Within a week I have nothing left, I'm done [*tchunado*]. Even sometimes, [my girlfriend] lends me money and then says that I'll have to pay her back double, as a joke, you know. So she has money but in terms of communications, she only sends *bips*!!! And then I phone her. With the money she lent me yesterday, I bought credit, but she sends me a *bip*, you see, so I phone, you see . . .

The gendered rules underpinning phone etiquette were so ingrained that Antonio's girlfriend, and Antonio himself, failed to adapt to real life situations. Determined to get his point across, Antonio added: "Or you meet a girl today and you give her your number. Then, the next day . . . she'll send you a *bip* and you have to respond. And to feel that you're a man . . . you just have to respond." Echoing Manuel's opening comment on the commodification of relationships, Antonio constructed a narrative in which masculinity had come to be increasingly gauged by one's ability to respond to *bips* and other requests for material support commonly placed via the phone. Among these, airtime was a popular gift that men gave to women and which was easily transferred between users in any denomination.

As a married and uneducated man, João's intimate experiences were somewhat different to those of suave and educated Antonio. João concurred with Antonio, but his resigned attitude contrasted with the latter's cynicism. For example, he said: "Sometimes a girl will ask you to send her 5 MZN [credit], now you, as a man, you won't say no." At the end of the day, both young men felt compelled, as men, to subsidize women's phone bills. Although both were self-conscious of spending their hard-earned money on women, they were nonetheless willing to play the game, since they saw bearing these costs not only as an expression of maleness, but also as a necessary feature of seduction.

In response to the two men's diatribe, Gina and Tereza agreed that they often used mobile communication to their advantage. For example, Gina said: "With my phone I can send a *bip* to someone and then I can get even more because, all of a sudden, that person can tell me: 'come get this thing,' a thing that I might have never even hoped for." They considered, however, that in reality, things were much more complicated and insisted that their experiences did not necessarily gel with the neat gender divide depicted by Antonio and João. To start, both argued that women could also end up, like men, "wasting" money on phone calls, and thus nullify the gains they may have made on other occasions. "Sometimes I top up to talk to someone about something very important," Gina explained, "but then I deviate [*desvio*] and I end up talking with other people. . . . Sometimes I regret it . . . [thinking to myself] 'shit, why didn't I call that person with whom I had something important to resolve?'" Gina and Tereza also argued that phone etiquette was often neglected and insisted that there were men out there who sent *bips* to women. "You know there are men who are conquered [*conquistado*] by women," Tereza explained, "and if the man finds out that a woman has a lot of money and that she is really fond of him, she'll say, 'Ok! Here you go.' Then he'll top up his phone but he'll still send you a *bip*, HIM as a man!" The two women agreed that these practices were considered "*feio*" ("ugly"; i.e., socially reprehensible) and that most people would be bewildered to learn that men could stoop so low as to send *bips* to women.

At this point in the conversation, Antonio interjected by saying that he recognized that there were men who did send *bips* to women but that these were exceptions that had little to do with more common experiences. He was arguably right. For the women, however, these exceptions were significant. In their eyes, they pointed to an increasingly fragile masculinity and more specifically to young men's inability to live up to mainstream ideals. Indeed, they insisted that it was men "nowadays" that contravened core social values and attributed this demise to men's own failures. By challenging the idea that patterns of phone use were gender-specific, the women were also questioning more profound ideas about female dependence and male autonomy. Like women, men also did send *bips* and, like men, women also did end up "deviating" and wasting credit on mundane conversation. And even more importantly, if male privilege rested, in part at least, on material grounds, what happened then when this material base was seen as crumbling away?

In defense of his earlier argument, Antonio added:

> Well, there has to be a lot of trust and many days that they are together [for a man to send his girlfriend a *bip*] because, it's like this: I know a girl today, you see, and she gives me her number, and then we start seeing each other. In the early days, I'll never have the courage to send her a *bip*, or else that girl is going to think that I am a loser [*matreco*]. But then once we are used to one another, then . . .

So, even Antonio sometimes sent *bips* to women. But that was not the point; rather, what mattered was that doing so was perceived as profoundly transgressive.

By testifying to a man's economic situation and owing to its role in seduction and in the management of intimate relationships, the ability to answer *bips* potently encapsulated mainstream ideas about youth masculinity in

Inhambane. As commentaries on social competence and gendered normative ideals, debates like this one were particularly revealing precisely because they highlighted how the introduction of a novel technology was participating, in conflicting ways, in crystalizing entrenched power relations while simultaneously offering a register to assess and sometimes challenge the very social hierarchies that underpinned these power relations. That is to say, phone etiquette acted as a new register to express the reconfiguration of gender relations and the redrawing of ideas of masculinity and femininity, while exposing the tensions that existed between gendered normative ideals and the socioeconomic constraints that young people were contending with.

Replying to *bips* was perhaps a small price to pay for these young men to regain some lost ground, given the slippery material base on which their claim to dominance had come to rest. For young men like João and Antonio who only had access to money sporadically, even the small gesture of replying to a *bip* was, at times, beyond their reach. *Biping* not only put pressure on these young men, but it also made their hardship more apparent. As João pointed out, "A girl will think: 'What can this man have to offer if he cannot even call me back?'" As such, young men were always on the *qui vive*, as they never knew when their masculinity would be probed next. Indeed, I often heard men describe their experiences with mobile phones as stressful.

In this context, masculinity went from being a social process that once took men to faraway places, to being understood as mercurial and hypersensitive to the vagaries of everyday life. By rendering these young men's economic difficulties more visible through unanswered *bips*, infractions to phone etiquette ultimately encouraged young women to question male dominance and female dependence, not so much to topple the system but rather to act on its ramifications. I therefore question whether phone practices were simply reaffirming patriarchal inflections, as has been argued elsewhere (cf. Batson-Savage 2007). Instead, I found that women had proven particularly agile at using the phone to their benefit. In fact, as discussed in detail below, women were often purposefully and quite knowingly taking advantage of the dissolution of patriarchal authority. In other words, if phone etiquette was informed by socioeconomic hierarchies and designed to make the better-offs pick up the tab, it, in turn, allowed young women to benefit by trading on their subordinate status.

Since the arrival of mobile phones, commitment to specific relationships—between friends, between lovers, or between an uncle and his nephew—has become quantifiable by how much one is willing to spend on phone calls and by how generous one is with transferring airtime. When Manuel suggested that relationships were commercialized, he was speaking of a profound crisis of authenticity (Archambault 2016). He was also hinting at the way some women have shrewdly started adopting a more selective, "materialist" approach in their choice of intimate partners. Ultimately, failing to reply to a *bip*, as the flouting of a basic expectation, translated into the dissolution of a man's claim to exclusivity, as it stood as a symbolic justification for women to turn to other men in order to answer their unmet needs.

In this economy, the better-offs were not only expected to bear the costs of phone calls, they were also asked to fulfill all sorts of requests that, as it were, were often placed via the phone. Apart from airtime, which was by far the most common, popular requests often entailed a sum of money for a specific purpose such as transport, photocopies, sanitary pads, food, medicine, and various other daily necessities. The phone, then, was not only a tool of persuasion, as discussed in the previous chapter, but also a powerful tool of redistribution. In the next section, I situate mobile phone requests within the wider redistributive economy before turning to the phone's integration in the sexual economy.

The Perfect Redistribution Tool

Several studies have shown how important mobile communication has become to the workings of redistribution networks (Horst and Miller 2006; Skuse and Cousins 2007; D. J. Smith 2006). In southern Mozambique where many have long relied, as detailed in chapter 2, on social networks to eke out a living, telecommunication has made it much easier to maintain and activate these networks.

During his 2013 visit at the University of Oxford, Mozambique's former President Joaquim Chissano told me a story that beautifully situates requests placed over the phone within a longer history of redistribution. Papa Chissano, as he is affectionately called, grew up in rural Gaza province[4] where most of the adult men worked on the mines in neighboring South Africa. At the time, husbands could only communicate with their wives through written correspondence until a public telephone was installed in the village. The women would gather by the phone box and wait for their husbands to call. When her turn came, one of the women, as the story goes, took the opportunity to tell her husband that their son had already grown out of his trousers and asked him to send them a new pair. As she gestured to show how much the child had grown—something people do all the time to mark height—she put the receiver down and inadvertently hung up the phone. Chissano described how

4. Gaza province is a province south of Inhambane province.

bemused the woman was by her husband's silence when she picked up the receiver again.

While I think the story was meant to show how rural people were a little clueless when it came to new technologies—and it reminded me of Kenneth's mother, who would always answer her newly acquired landline with the receiver upside down as she was convinced that the cord should be coming out of the top—it highlighted important continuities regarding communication technologies and redistribution.

As Keith Breckenridge (2006) showed, throughout the twentieth century, "letters served as one of the few tools those who remained at home might use to extract resources from the small wage workers received in the towns" (146). The majority of letters exchanged between migrant workers and their families in southern Africa, like the phone calls Chissano recalled, were concerned with "household management" (ibid.; Schapera 1941). In addition to real-time communication, mobile communication has enabled those at the receiving end to play a more active role in the redistribution process, like letters once did (Breckenridge 2006), only this time, in a far more efficient and accessible manner. In Jhoker's words, "Before you had to wait for your uncle to remember to send you money for school, now you can call him to remind him."

"I call for a better life" (*Chamo para uma vida melhor*) was what Manuel, a young man in Inhassoro,[5] one of the districts in the province of Inhambane, answered when I asked him what he used his phone for. To illustrate his statement, he recounted how he had managed to convince his uncle who was working in South Africa to send him a bicycle. "I just kept sending him *bips*," he said, "and after some time, he admired my determination so he sent me a bicycle!" Once coordinated via the phone, remittances, whether in cash or kind, were then usually delivered through more "traditional" channels, by family members themselves, by acquaintances, or by young men working on public transport (*cobradores*) (de Vletter 2006: 20).

When I first met Saulinha, she was a little embarrassed to show me her low-range Nokia handset. "It's a cheap model," she explained, "but I just need a phone to communicate, you see, I'm studying here, but all my family lives in Maputo and I really depend on them. . . . The other day, I needed money so I sent a *bip* to my father but he didn't reply. Then a friend of mine came by and she had phone credit, but she didn't have a phone, so we exchanged services:

5. The following examples all come from a brief period of field research I carried out in the district of Inhassoro where more people relied on remittances than those in the city of Inhambane.

I lent her my phone and ended up with the free messages.[6] . . . So I sent a text message to my father asking him for 70 MZN. Then my father sent me the money with a minibus driver." As these two examples illustrate, telecommunication has played an important role in the redistribution process, and as the following one indicates, the few who did not own a phone risked being seriously marginalized. Gito lost both his parents during the war and at the age of nineteen, he was struggling to get through secondary school. Unable to afford a phone, he would buy credit and look for a friend or neighbor willing to lend their phone. "Sometimes I call my father's family in Maputo for help with school [fees and material]. My survival depends on begging, what can I say? But I wish I had my own phone! That way they could also contact me and tell me if they can send me some help." Young people in Inhambane relied on social networks built on kinship, friendship, or sexual relationships to facilitate their access to money and favors, including state resources. Securing a passing grade, finding a job, and getting into university, like obtaining prompt treatment at the hospital, usually depends on one's connections.[7]

In the chapter on petty crime, I showed how young men develop social networks in which they position themselves as brokers to peddle stolen goods. In what follows, I turn to the sexual networks young women cultivate in their efforts to negotiate everyday uncertainty and to create fulfilling lives. I preface the discussion with two highly popular pop songs by the singer and songwriter Denny OG, which offer colorful, but also cynical, social commentaries on the commodification of sexuality through the idiom of consumption.

"*Macarapão*" (Mackerel) recounts the story of women who are forced to sell mackerel (i.e., sex) to get by. The song suggests that it is due to poverty and the lack of opportunity that women turn to transactional sexual relations in order to get by. In the song, one woman asks rhetorically: "What am I to do given that I don't have money? When I ask for a job, the boss says, 'girl, go to school.' . . . That's why I sell mackerel!" These themes are picked up in a second song, entitled "*Mabatata*" (Potatoes), which highlights the scope of the phenomenon. To make his point, the singer enumerates all the women he knows who sell potatoes (i.e., sex): Amelia, Tereza, Sabina, a doctor's wife, even his own wife. Again, poverty is identified as the main driver behind the sexual economy. One man then asks his wife: "Don't you feel ashamed to be sitting at home? Are you not aware that the price of bread has just risen?" The

6. Every phone top-up comes with a number of free text messages.

7. In her research on Internet use among youth in Accra, Ghana, Burrell (2009) similarly shows how users privilege the development of personal contacts over "depersonalized information" (158).

woman replies that she is "selling potatoes" to help her husband. Powerful for its irony, this commentary offers an unusual perspective on intimate relationships as popular music more generally explores cases of deceit and jealousy. Here, food metaphors speak of how men consume women sexually but also about how women themselves consume men financially. Like most of Denny OG's songs, these two are social critiques that arguably act as "maps of experience" in ways similar to the *paiva* song a couple of generations earlier (cf. Vail and White 1991), even if, by the singer's own account, songs are often misinterpreted as endorsing certain practices rather than as interrogating them.[8]

Chular and the Workings of the Intimate Economy

Sexuality in sub-Saharan Africa, and in southern Africa in particular, has attracted much attention, especially since the onset of the HIV/AIDS pandemic. There is a large body of literature on sexual networking that examines the political economy of sex in the region and that makes important contributions to our understanding of livelihoods on the one hand, and of intimacy on the other.[9] One of this scholarship's main contributions has been to highlight how women often participate in relationships built on the exchange of sexual favors for material gain as active participants rather than as passive victims of predatory men. To quote a report on transactional and transgenerational sexual relations set in Maputo: "Young women learn that their sexuality is a valued resource (as a result of structural conditions and balance of gender and power relations), and exercise agency to gain financial resources from older men for sexual services, often with multiple partners to maximize the benefits" (Hawkins, Mussá, and Abuxahama 2005: 3).

What this scholarship also offers is an ethnographically grounded discussion that situates the exchange of sex and money within intimate geographies in which such exchanges foster rather than undermine intimacy. An emphasis on female agency should, indeed, not overshadow the crafting of new and meaningful intimacies and the importance of experiences with love and affection (Spronk 2009; Thomas and Cole 2009: 10).

I am interested, here, in the ways in which women in Inhambane have harnessed mobile phones to better tap into the intimate economy. But first, a few words on *chular*. *Chular* means taking advantage of someone under

8. Interview with the singer and songwriter Denny OG, Inhambane, April 28, 2007.

9. For regional studies, see Hunter 2002; Silberschmidt 2004; Stark 2013; Thomas and Cole 2009; Thornton 2008. For Mozambican studies, see Arnfred 2011; Bagnol and Chamo 2003; Groes-Green 2010; Hawkins, Mussá, and Abuxahama 2005; Manuel 2008; Pfeiffer 2004.

sexual pretenses and usually under the guise of a relationship, which may involve varying degrees of emotional attachment. Extractive but also affective, to varying degrees, it is generally something women do to men. There are important similarities between *chular* and the basic premise of transactional sex, but *chular* is also different in that the terms of the exchange are more ambiguous and the outcomes generally more uncertain. It does, however, rest on a similar sexual ethos, according to which money, sex, and affection are entangled in a complex web of expectations and obligations.

Women's access to resources in *chular* relationships rests on the promise of an exchange of sexual services that may or may not materialize, as reciprocity in *chular* exchanges is sometimes deferred, sometimes written off all together. When women speak of having "*chular*ed" a man, they usually mean that they managed to convince him to part with money—either in the form of drinks, food, or cash—without having given anything substantial in exchange.

Some of the young men in Liberdade claimed they were too clever to fall for women's maneuverings, but, in my opinion, that was often essentially because they did not have money to part with in the first place. In fact, whenever they did come into money, these same young men would generally change their tune and start spending on women. Although their economic marginalization did not necessarily translate into abstinence, the young men I worked with repeatedly complained that they lacked the resources needed to play the game and that they struggled to attract young women who, as they argued, only had eyes for older men with money. As Antonio once lamented, "There was a time when you could get a girl for a piece of candy, but girls these days want flashy mobile phones and designer clothes. They only have eyes for older men [who can offer them all these things], and us youths, we just can't compete." *Chular* exchanges were, in fact, commonly transgenerational: they brought together young women and older men, and sometimes young men and older women.[10] If access to income has historically been unevenly distributed along gender lines and to the advantage of men, such gendered privileged access to income has increasingly become the preserve of a few, generally older men.

Although *chular* rests on the exchange of sexual favors for material gain, at least in theory, it is seen as radically different from prostitution. For the woman I worked with, *chular* is just a thing that women do to men. Prosti-

10. My findings are in many ways similar to the ones presented by Bagnol and Chamo (2003) on intergenerational sexual relations in Zambezia, a province in central Mozambique, as well as to those reported in other parts of the continent, such as in Nigeria where young women also use their phones to negotiate their relationships with "Sugar Daddies" (D. J. Smith 2006).

tution, on the other hand, is understood as qualitatively different, as involving a fixed price for a specific service and none of the other aspects of intimacy such as affection, company, and the opportunity to *desabafar*. In other intimate exchanges, in contrast, when men give money to women they generally make sure to construct it as money for a specific purpose: money for breakfast, money to pay for photocopies at school, money to buy medicine for an ailing relative, or more generically, money for bread. Even though the money may very well end up being used for anything but the stated purpose, the idiom of the gift avoids it being construed as a form of payment.

The line is, however, a fine one.

Marta once used a market idiom to describe how she was getting by. "I'm eating men's money," she explained. "I charge them!" When I pushed her further on this, Marta soon realized that I was getting the wrong impression. "Well I'm not really charging [these men], you know?" she added, fearing that I might misconstrue her activities as prostitution. She clarified:

> Nowadays, girls go to Maxixe to prostitute themselves, then they come back to Inhambane, as if they were all innocent, they even look at their parents straight in the eye. I would never do that. Women have a lot of power. They have something men want, but if you sell your body, then even if you end up marrying one day, your husband will find out about your past and then he'll never respect you.

Marta felt there was an important difference between prostitution and her own involvement in *chular* relationships. She told me about one of her neighbors "who really actually charges." She explained: "She even takes phone calls and meets up with men for sex." But she added, in contrast: "I have relationships with men that I like, unlike the prostitutes who have to go with whoever comes along." *Chular*, unlike prostitution, involves choice and an open-ended negotiated encounter.

The difference between *chular* exchanges and prostitution was also meaningful to the young men I worked with. Soon after moving to Maputo to work as a security guard, Helder started being approached by young women, intent on seducing the new guy in town who, as rumor had it, had a decent job. He accepted to exchange numbers with some of them but made it clear that he had no time for a girlfriend. He had just started working and wanted to concentrate on "building his future," as he put it. If he accepted to give out his number, he said it was because he was weary of making enemies in a place where he hardly knew anyone. Already disillusioned by women, Helder did not flatter himself when some of them insisted on getting in touch by sending him "endless *bips*." He showed me a message that one of them had sent him, adding that he had saved it so that I could see for myself how women could be *interesseira* (having ulterior motives). In her message, the young woman explained that she needed 250 MZN to purchase medicine for her daughter who had been ill for over a week. She concluded the message by saying that she'd be happy to come around to his house to collect the money later that evening. Helder was shocked at the woman's forthright manners. "The girl is actually offering to sell herself for the night! She even has a set price! I wonder how many other men she's sent this message to," he said with disgust.

Back in Inhambane, a young woman pointed out: "*Chular* is a game. If you don't want to give sex, you need to know how to talk to close the man's eyes, all you need is the right *papo* . . . all you need is *visão*." Owning a mobile phone also helps when playing the *chular* game.

To convince me of how easy it was to get things "with her phone," Mimi once gave the following demonstration. She started by sending a *bip* to one of her suitors. A few seconds later, the man phoned back, giving Mimi the opportunity to complain about being out of bread. Within an hour, the man was at the door with two loaves of bread. Not only had he paid for the bread but he had also covered the cost of the call asking for the bread. On another occasion, Mimi kindly shared details about the various contacts she had stored on her phone. After going through them alphabetically, my tally revealed that she was intimately involved with over ten different men. She smiled and asked how I thought she had managed to build a house with a corrugated iron roof. A few days before Christmas, Mimi purchased airtime and called most of the male contacts she had stored in her phone. The phone calls I overheard were all very similar. Mimi would start by saying that she was phoning to ask for her Christmas present and would then offer to pick it up herself. She was persuasive, almost threatening, and it was clear that whoever refused to comply with her demands was jeopardizing his chances of ever getting close to her. Not only was the phone enabling young women like Mimi to be more unrelenting in their requests to lovers and suitors, it also allowed them to do so discreetly and at a distance. In such technologically mediated exchanges, young women could sometimes even further subvert chains of reciprocity by receiving without giving.

Some of the young women I knew were particularly proud to share their tricks with me. For instance, Anita would make her way to one of the city's bars, usually with a girlfriend, in search of a man willing to buy her and her friend drinks and dinner; in other words, a man willing to subsidize their evening. She would sometimes end up leaving with the man in question, but more often than not, she would head back on her own, sometimes in my company. And when the man proved a bit too insistent, Anita knew exactly how to plot her escape through the meandering alleys that separate the city from the suburbs. Anita would pretend to go to the toilets and make a quick escape through the back. It usually worked, she assured me.[11] Many of the young women who frequented such establishments were, like Anita, from the suburbs, areas that would have been unknown to the male patrons, many of whom were either based in the city center and often from out of town, such as civil servants, tourists, and students from Maputo who were attending the local university. If she ever crossed paths again with these men, Anita would make up an excuse about herself or her friend having fallen ill at the time. But Anita did not, in any case, feel indebted; instead she insisted that she had already given back simply by accepting to keep the man in question company throughout the evening. Reciprocity, in these intimate encounters, is far more subtle than a simple exchange of sex for money. Accepting gifts—even beer and a serving of barbecued chicken and chips—from a man is a relational act and for some men, this sort of recognition is sometimes good enough.

Either way, men were, on the whole, and as hinted at earlier in this chapter, often playfully cynical about the rules of engagement in the intimate economy. For example, Mikas, an employed and older man, often complained about women taking advantage of him, of "enslaving" him, as he called it. One day in my company, he received a phone call from a young woman he had been pursuing for some time. Already imagining the motive of her call, he answered on loud speaker for my benefit. The woman went straight to the point and asked

11. Throughout my time in Mozambique, I heard young men tirelessly comment on their insatiable sexual appetite. I was convinced that everyone had incredibly active sex lives. As I dug deeper into this intimate economy, however, I found that people were, in fact, having relatively little sex. When I eventually started asking for details—when, for example, I asked some of my male research participants how often they had sex, how many sexual partners they had, and when the last time was that they had had sex—these same men had to concede that, in reality, they only had sex once in a while, sometimes less than once a month. There were several factors that worked to the advantage of women wishing to take without giving and that made having sex in this small town logistically complicated. For one thing, it was considered transgressive to have sex at a woman's house, and although people did make exceptions every now and then, women could easily deny men access to their sleeping quarters. Traditionally, a man who had sex at a woman's house would then have to sacrifice an animal and perform a cleansing ceremony, and although I never heard of anyone actually performing this ritual, the thought of having to do so nonetheless acted as a powerful deterrent. Married men, for their part, were in no position to bring other women home. They could rent a room or ask a friend or brother to lend them one—for a while, a couple living near one of the local *baracas* had a tent up in their yard that they sometimes rented out to couples looking for a bit of privacy—but women were easily put off by these options. In short, if the feeling was not quite mutual, women could argue themselves out of having sex without necessarily having to flee.

him for phone credit. After hanging up, Mikas started laughing: "I'm a fool, I'm stupid. I'll give her credit and then she'll use it to phone another guy!" He then told me about another young woman who had recently contacted him for assistance after claiming that someone had stolen all her clothes. Mikas explained that he gave her money to purchase new clothes, along with 80 MZN top-up so that she could phone him and keep him updated. But she never phoned back. Mikas knew he had been duped and he added, with a touch of irony, more than bitterness: "Whenever I phone her, she is either at church or on her way to school!" Men who fell "victim" to requests placed via the phone were mostly older, employed men like Mikas, but younger men were nevertheless also regularly asked to reply to *bips* as well as to supply smaller things like hair extensions, lunch money, or phone credit. Men may have always competed among themselves but with women's demands and expectations increasing, and given the scarcity of reliable sources of income, they had to spread themselves ever so thin.

There is a large Africanist literature that shows how redistribution is often driven, in part at least, by the fear of occult reprisal, leading, in turn, to what Eric Gable (1997: 215) aptly calls "nightmare egalitarianism." Generosity can, in a sense, be understood as a prophylactic against witchcraft. Inhambane residents, like others on the continent, also commonly use the "idiom of consumption" to articulate their views on, and experiences with, inequality (West 2005: 37; see also Bayart 1989). Songs and proverbs stand as constant reminders of the uncertain nature of prosperity and, more implicitly, of the risks of sorcery that eating alone potentially attracts. As Inhambane residents constantly remind each other, "refusing is ugly" (*negar é feio*) (see also West 2005: 37). I would argue, however, that those who gave were not only guided by such pragmatism. Although those who had access to a reliable source of income often spoke of the stifling pressures to redistribute, they shared not only because they felt compelled to do so, but also, in a sense, because they could. By giving and taking, individuals were able to negotiate the positions they occupied in these networks and craft fulfilling lives through various relations of dependence and interdependence. As Parker Shipton (2007) notes in his study of entrustment among the Luo, "what one borrows or lends helps define who one is" (14). Sasha Newell (2012) similarly shows how the potlatch-style display of resources and conspicuous consumption so central to the youth culture in Abidjan highlights the social value of social relations over profit and accumulation. He writes: "Of course, in some more abstract sense, people profit from their social relationships, but the point is that the social relationships take priority, or rather, that the maintenance and accumulation of these relationships is its own kind of profit" (66). In

Inhambane, the men who allowed women to "*chular* them," so to speak, were also driven by the profound desire to feel relevant, "to feel like somebody" (*se sentir alguém*). By spending time with these men, I gained an understanding of how it made them feel to treat women to food and drink even if, and on their own admission, these women were arguably taking advantage of them. Giving felt good. Period. Although it did feel even better when it then came with a bit of loving. The line between seduction and predation is, indeed, often a fine one and there is often no clear victim or perpetrator, no indisputable winner or loser. Playing the *chular* game procured excitement, access to new experiences, and a sense of being somebody; of being alive! In the Inhambane context, where opportunities are so very limited, where desires, aspirations, and dreams are so easily thwarted, being able to do something is, well, quite something. Likewise, answering *bips* was about demonstrating maleness but it was also about feeling like a man and, ultimately, about feeling alive, at least every now and then. As we saw earlier, however, sometimes even the simple act of replying to a *bip* was beyond their means. It was usually, then, when they found themselves unable to participate in the intimate economy that young men complained about the commodification of intimacy.

The Commodification of Intimacy and the Crisis of Authenticity

The commodification of intimacy is part commentary about rising inequality, part commentary about gender and generational hierarchies (Hunter 2002; Thomas and Cole 2009). Although young men did not challenge the ideal of the man as provider—providing for women was seen as undeniably part of being a man—what they found problematic was that the material component had become so central so as to supersede, instead of participating in the crafting of, genuine intimacy. They liked to paraphrase a classic Portuguese text taught in the primary curriculum entitled "Nowadays there are no longer women to marry" (Sebastião 1999: 50). There are women (*mulheres*)," I was told, "but no woman (*mulher*)." The men accused the women of being "materialist," "corrupted," and privileging material gain at the expense of "true feelings" (*sentimentos verdadeiros*). In doing so, they conflated women's perceived transgression of two competing ideals of femininity: on the one hand, the ideal of the "respectable African woman" suitable to become a wife and mother; and, on the other, the "foreign-inspired girlfriend" with whom one imagines building a relationship based on true love and aspirations of exclusivity, resembling Giddens's (1992) "pure relationship"—an intimate relationship built on equality and mutual disclosure. Young men were nostalgic of a not-so-distant past, before women became "materialistic," when marriage was believed to be a genuine and lasting union between two individuals and their families, if not one always initially based on love. They were also nostalgic of a genuine intimacy that might have been, one built on feelings of true love between a man and a woman—these are undeniably heteronormative ideals—rather than on negotiated agreements between male elders. Young people found food for thought on intimacy in Brazilian telenovelas, Pentecostal sermons, NGO slogans, party politics, and everyday dealings with tourists and expatriates.

It has been argued that the commodification of intimacy need not necessarily translate into a diluted or counterfeit form of intimacy. As Nicole Constable (2009: 58) notes in a recent review: "Globalization does not simply result in greater commodification of intimate sexual, marital, and reproductive relationships; it also offers opportunities for defining new sorts of relationships and for redefining spaces, meanings, and expressions of intimacy that can transform and transgress conventional gendered spaces and norms." In *Temporarily Yours*, for instance, which charts recent transformations in the meaning and marketing of sex work in San Francisco since the 1990s, Elizabeth Bernstein (2007) challenges the idea that the commodification of sexuality has resulted in the dissolution of intimacy by showing instead how the current sex market offers a space for the procurement of "authentic, yet bounded, forms of interpersonal connection" through what is known in the industry as the girlfriend experience (21). Indeed, commodification unsettles intimacy in various ways and, as Peter Geschiere (2013) writes, "Commodification may seem to be opposed to intimacy, but in practice it seems to trigger a search for new intimacies" (67). Debates around the commodification of intimacy in this part of the world have a long history. As Thomas and Cole (2009) write, "Africans have long forged intimate attachments through exchange relationships. They have also long grappled with the ways in which monetization strains this practice" (23). Throughout the region, we find historical struggles that pit men and women, young and old, against each other, in ways that speak of contested visions of modernity (Ivaska 2011).[12]

For the men in Inhambane who were too poor to "to afford a girlfriend," as some of them put it, the commodification of intimacy certainly felt momentous. And although the gender and generational cleavages that it seemed to amplify were the product of a long and complex history (Lubkemann 2007;

12. Masquelier (2005: 77) quotes a Hausa poem written in the early 1970s, which describes women as corrupted by money.

Webster 1975), the enthusiastic uptake of mobile phones was clearly identified as a catalyst in that it allowed women to be ever more relentless in their requests. In *Married but Available*, a novel set in a fictitious African country, one of Francis Nyamnjoh's (2008) characters explains: "For young girls, they are mostly using the mobile phone as a tool to grab things left and right, and also, to make themselves available for grabbing. When a woman gives you her phone number, she is actually giving you access to herself, and also as a way to pester you to send them airtime, this and that" (126). The phone, in other words, as a tool of persuasion, was easily turned into a pestering tool.

If the debates sparked by the introduction of mobile phones speak of perennial moral concerns, it is nonetheless also important to highlight how, alongside such continuities, there is also a significant redefinition of what intimacy could and should mean. Instead of precluding intimacy, the commodification of intimacy has also encouraged the pursuit of alternative forms of intimacy (Archambault 2016). In this intimate economy in which the main losers were young men and older women, as the former lacked the financial capital of older men, and the latter lacked the "bodily capital" (Wacquant 2004) of younger women, it had also become increasingly common for young men to engage in intimate relationships with older women. Older women with money, that is. This, however, was still an emerging trend and one that was extremely frowned upon. Boys are told, while growing up, that they must never have sexual relations with older women, especially not with widows, and are warned that acidity of older women would burn their penis. Still, an increasing number of young men are taking their chances.

Sitting at a neighborhood *baraca* one Sunday afternoon, Lulu tried to convince his friends that they had everything to gain from dating older women. "They're nice," he explained. "They give you more affection, as if you were their son. One just bought me a phone." Lulu equated affection with gifting in no uncertain terms, yet by comparing his relationship with an older lady to the one between a mother and her son, he was also emphasizing the nurturing dimension of the relationship. Though no one would have denied the calculative dimension of such asymmetrical relationships, this did not necessarily preclude the possibility of an affective connection. These older women were, for their part, also asserting themselves against the older men who were often more interested in spending their time with, and money on, younger women.

In these stories of pretend-love, infidelity, commodification, exclusion, and inversion, the phone often played a central role. It was because of the phone that people argued and broke up, and it was thanks to the phone that people cheated on each other and that women had turned exchange into predation.

Indeed, the phone offered more than a useful register to express and debate the redefinition of intimacy underway in Mozambique; it was also understood to drive and enhance these transformations. Yet, what Mozambicans also made clear was how individuals were using mobile phones to help cover-up and silence social contradictions and, in doing so, succeeded in preserving an "ugly" public secret about the postsocialist, postwar intimate economy.

6

Truth and Willful Blindness

If absolute certainty is an illusion (Wittgenstein [1969] 1975), then clarity might be a reasonable compromise. With accurate information, one can take informed decisions and count on predictable outcomes. One can treat, mitigate, prevent, remedy, invest, or opt for inaction, all with some sense of direction. Often, the search for clarity involves enlisting the help of specialists in the art of disambiguation—diviners, doctors, intellectuals, scientists, journalists, investigators—and relying on various technologies and techniques. Divination, a privileged topic of anthropological inquiry, has been understood as an effort to generate truths about the world and make visible the invisible or the unknown (Shaw 2002), ever since Evans-Pritchard (1976) showed how the Azande relied on divination through *benge*, the poison oracle, to identify the witches in their midst. But the quest for clarity can also operate at the less esoteric level of vision and visibility. For example, Maurice Bloch (1995) shows how Zafimaniry concepts of clarity inform their understandings of deforestation in a place where a clear view is a good view.

In *Ethnographic Sorcery*, Harry West (2007) draws interesting parallels between sorcery and ethnography and highlights how both can be understood as efforts to go beyond appearances and to "gain interpretative ascendency in and over the world" (80). West also suggests that it is through such processes of interpretation that we—sorcerers and anthropologists, as well as other interpreters—make the world. Indeed, for nearly a century, anthropology, which emerged and took shape through encounters with alterity, endeavored to explain sociocultural variability. From functionalism to structuralism to symbolic anthropology, and passing by Steward's cultural ecology and Ruth Benedict's culture and personality school, anthropology remained committed to the task of uncovering what lay beneath the surface of social practices and cultural forms, whether its focus was on functions, structures, symbols, areas, or patterns. Mainstream contemporary anthropology is still very much invested in disclosure, although the things we hope to uncover are no longer functions and structures but historically contingent power relations (Ortner 2006), deeper meanings (Ferme 2001), assemblages (Ong and Collier 2005), and ethics (Lambek 2010).

In this search for clarity, we rely on a careful balance between attention to vernacular experiences and articulations on the one hand, and heuristic skepticism on the other. Anthropologists are trained to take seriously, rather than dismiss, even the most obscure propositions, but, at the same time, to refrain from taking anything, even the most straightforward of propositions, at face value. Ethnography, in a sense, is the art of juggling this tension between skepticism and acceptance.

In recent years, however, some of the proponents of the so-called ontological turn have, in their critique of representation, recalibrated their attention toward the surface of things, toward the literal and away from the metaphorical (Candea 2011; de la Cadena 2010). I have engaged with this suspension of skepticism already (Archambault 2016), but here I wish to interrogate the anthropological project of uncovering from another angle, to reflect instead on what happens to this project when we come across ignorance as social practice.

It is only recently that anthropologists have started investigating everyday modes of not knowing. As Wenzel Geissler (2013) points out, "unknowing" has rarely been examined as "a creative social form in its own right" (15). Instead, ignorance has mainly been understood as the passive reverse of knowledge (Mair, Kelly, and High 2012).[1] As a constituent part of any regime of truth, ignorance is entangled in practices of secrecy; in fact, if secrecy involves the control of information, then ignorance is often an outcome of secrecy. The kind of ignorance I am interested in here is, however, often contrived or at least more complicit. In this chapter, my reflections depart from the more familiar social-constructivist debates around relativism and the constructed nature of claims to truth, to focus instead on relational truths, that is, on everyday propositions—my boyfriend is cheating on me; my girlfriend is faithful; my daughter is a *moça de casa* (a good girl); the neighbor's daughter is a *magapunza*[2] (a girl who sleeps around)—which may either be ignored, challenged, qualified, or accepted as truths.

1. An earlier version of the argument presented in this chapter as well as some of the material on which it builds initially appeared in Archambault 2013.

2. Literally, it means a woman (*maga*, a prefix that designates women) who likes to roll around (*puluwunza* in Chitswa and Gitonga).

Throughout this book, I have been arguing that the phone operates as part of a wider arsenal of pretense designed to silence and conceal dissonances and have shown how the phone—mobile communication, to be precise—opens up intimate spaces within which illicit pursuits can be negotiated with some degree of discretion. In this chapter, I push this idea further by showing how mobile communication helps preserve an unpleasant public secret about the workings of the intimate economy in which young women are encouraged to multiply intimate relationships and exchange sexual favors—or at least the pretense of such favors—for material gain on an unprecedented scale. I argue that the phone helps reproduce epistemologies of ignorance—modes of not knowing—by allowing everyone to pretend not to know. In fact, even the intimate conflicts fueled by intercepted calls and messages (chapter 4) further illustrate how intimate conflicts emerge when the phone fails as a tool of pretense and fires back. Such shortcomings are precisely what Mozambicans hint at when they describe themselves as beginners. With time, the assumption is, they will learn how to circumvent such conflicts. The various rules and strategies espoused to avoid intimate conflicts are, indeed, specifically designed to override the phone's revelatory powers in a regime of truth in which what counts as true is inextricably linked to visibility and, to some extent, also to audibility. Provided there is a real attempt at concealment, provided something—an affair, for example—is not "shown," its existence can be denied by the deceiver and overlooked by the deceived. In the following section, I start by situating willful blindness in contrast to the search for clarity and draw attention to the social value of open-endedness.

The Search for Clarity

When faced with misfortune, the youth of Liberdade often enlisted the help of specialists to uncover the root causes of their troubles and to gather advice on how to mitigate the situation and move forward. This could entail visiting a traditional healer, a pastor, the hospital, or the neighborhood secretary. Most of the young people I worked with insisted, however, that they were not really involved with "tradition" (*tradição*), the generic term inherited from the Portuguese and later rigidified during the fight against obscurantism waged by the socialist regime in the 1980s, which was used as a blanket term for all things occult, including divination, traditional medicine, and witchcraft. Inspired by Samora Machel's vision of the "New Man," these young people abided by modern, civilized ways of engaging with "tradition." Unlike in other contexts where the rejection of tradition has been described as an ongoing battle often waged with the help of Pentecostal and other charismatic churches, in Inhambane the rejection of tradition also spoke of a very real interruption in the transmission of so-called traditional knowledge following not only Frelimo's efforts at eradication but also as a consequence of the very tangible material constraints of the war years and the premature death of knowledge holders. These so-called traditional means of understanding and engaging with the world operated alongside competing systems of knowledge, and even though young people attempted to distance themselves from "tradition," it nonetheless informed local understandings of how to uncover *the* truth. Some were particularly cynical; all were eternal sceptics. Many were also astute political observers profoundly aware of the machinations of the state and of other more nefarious forces.[3]

Most of the time, however, these young people relied on themselves in their efforts to gain "interpretative ascendency" (West 2005: 80). They were, in their everyday meanderings, always on the lookout for snippets of information—they read footprints, expressions, demeanors, comments seemingly made in passing—that would provide insight into incidents, intentions, and collusions. *Visão* is, indeed, an everyday technique of clarity that enables one to see through appearances. Someone with *visão* knows how to navigate everyday uncertainty and make the most of "a state of limited resources for action" (Whyte 2009: 214). Someone with *visão* also knows how to conceal their tracks. *Visão* is precisely the ability to capitalize on seeing while toying with the visions of others to evade detection. It is, first and foremost, an interpretative endeavor. But someone with *visão* also knows how and when to look away and feign ignorance. Like being known, which puts one in a position of vulnerability (Comaroff and Comaroff 2012: 60), so can knowing be detrimental, all the more so when knowing calls for necessary but undesired remedial action. People often made an effort, especially when it came to matters of the heart, to remain in the dark and cultivate ignorance, not as a despondent response to unknowability but rather as a desired outcome. We saw, for example, how anger following the interception of a compromising call or text message was often directed at the phone for foreclosing ignorance as a possible option. In other words, uncertainty is often sought after and sustained, if only to justify inaction—in other words, "willful blindness." As I show below, the discretion granted by mobile communication, even if imperfect, has allowed everyone to keep face without having to address an unsettling reality.

3. Some were readers of the independent online newspaper aptly called *@Verdade* (The Truth).

Intimate life in Inhambane is, to borrow Berthomé, Bonhomme, and Delaplace's (2012) formulation, "predicated on opacity" (181). As much as they were determined to see beyond the surface, to transcend appearances, Inhambane youth sometimes opted out of a deeper knowledge, sometimes preferred uncertainty over clarity. Indeed, while successful living rests on the careful manipulation of social networks, it also depends as much on what one sees as on what one is able to overlook. Alongside practices of hypervigilance designed to uncover the truth as a way of dealing with uncertainty are practices of concealment that, in turn, perpetuate in productive and sometimes destructive ways epistemological uncertainty. The known and the unknown, then, can be turned into resources for action and, often just as importantly, for inaction.

The conundrum is simple: women's involvement in the intimate economy is profoundly transgressive but it also makes life a little sweeter,[4] not only for the young women themselves but also for their children, parents, and siblings and sometimes even for their other partners, by encouraging the redistribution of unequally distributed resources. But, as the Cabral snapshot in chapter 2 made clear, for everyone to coolly sit around the table, there has to have been an effort at concealing the source of the "nice breakfast." Depending on the circumstances, it may therefore be preferable for those indirectly involved to turn feign ignorance. And this is also true of the young men who are unable to provide for these women. It may very well be in their interest to turn a blind eye when their girlfriend goes after another man to secure what they, themselves, cannot deliver. As a tool of dissimulation, the phone helps preserve what Michael Taussig (1999) would describe as "a great 'as if' " (7), given that everyone sells potatoes, even the doctor's wife!

Control[5]

The world beyond the confines of the yard is fraught with risk and temptation, and women's movements must therefore be controlled for these risks to be avoided, or at least mitigated.[6] We saw earlier that a good woman is one who spends most of her time at home and who justifies her comings and goings, whereas a woman who is always seen out and about risks being described as

4. A euphemism commonly used is "sugar." For example, at a public meeting chaired by the Secretary of Liberdade, residents were urged to "ask questions whenever [their] daughters came home with sugar."

5. An earlier version of this argument was published in *American Ethnologist* in 2013.

6. Men, for their part, are relatively free to roam around as they please, though the married ones risk being met by an angry wife if they venture home too late.

vadia (vagabond) or a *moça da rua* (street girl). Closely tied to female sexuality and reproduction, the restriction of women's freedom of movement has long been a preoccupation in the region, which was widely debated in songs and other oral performances at least as far back as the early twentieth century (Junod [1912] 1966; Vail and White 1991: 125). Several of these songs detail the distinct challenges that men faced in rural areas involved in the migrant labor economy. Given that labor migration often kept men away from home for the better part of the year, married women were usually left under the supervision of their husbands' relatives. The patrilocal household was, indeed, usually characterized by fluid cohabitation arrangements whereby a household often included members connected through alternative networks of friendship or fosterage (Webster 1975). Following widespread displacement and wartime casualties, the patrilineal and patrilocal household historically found across the region was further transformed in the 1990s (Chingono 1996: 220), and, more recently, by people's search for education and employment opportunities in towns and cities. In Liberdade, the importance of women-headed households is the most visible manifestation of these changes. In such households, the elder son usually acts as the authority figure, but as most lack the means to back up the position, "fatherless young girls" (Arnfred 2011: 94) generally enjoy greater freedom than those living with a father. Remember the Cabral brothers' unsuccessful attempts at preventing their sister to go out at night. The men who worried about their women's loose mores were not only brothers but also boyfriends, husbands, fathers, and uncles, and they all had interesting things to say about the reasons why they found it so difficult to keep their female kin in check.

Generalized secondary school attendance has also provided young women with increased freedom of movement and, although it is still too early to determine how education will shape women's access to formal employment, the "side effects" of going to school every day are more readily observable. Temporarily freed from the surveillance of male kin, young women can now encounter new opportunities "along the way" (*no caminho*). Echoing earlier songs sung by worried migrant workers, several pop songs offer damning commentaries on older men who try to seduce school girls on their way to school by offering them a lift or something to eat.

Another vivid image often included in discussions on moral decay is that of young women escaping from the window at night. The image is a powerful metaphor, especially given that very few houses in Liberdade actually have windows, that exonerates parents through an emphasis on these young women's agency. Parents insisted that they controlled the door, the exit used by those with nothing to hide, but that there was little they could do to prevent

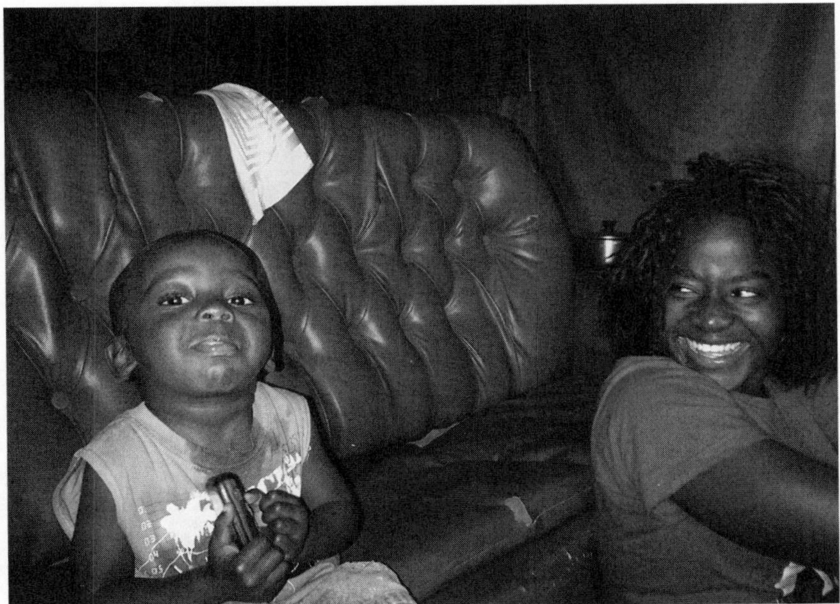

FIGURE 10. Taninha with her son, Inhambane, 2012. Photo by author.

secretive nocturnal excursions through the window. The image of women escaping from the window at night can be understood as a variation of the idea of jumping the fence found in other parts of southern Africa (see van Dijk 2012: 144) that similarly shifts the blame on the cunningness of the absconder and away from the carelessness of the person who is responsible for keeping household members in check.

Women, in their defense, argued that they were the victims of male machination and deceit. The growing number of children born out of wedlock was, in fact, understood to reflect much broader societal moral decay, one in which men were failing to take responsibility for their own actions. "Men, these days, refuse to take responsibility!" I heard young and old repeat so often that I was almost convinced. As I came to better understand the intimate economy, however, I developed a qualified understanding of this pervasive claim. It was undeniably the case that men who impregnated young women out of wedlock escaped with little consequences other than having to pay a fine to the girl's family, especially if the woman in question was fatherless. Still, women were not always the innocent victims that they were made out to be. The following examples will make this clear.

When I met Amelia, she was living with her two children at her sister's house, and when I asked her where the fathers of her sons were she answered with the classic refrain. However, during a later conversation, Amelia admitted that the father of her first child had come to introduce himself to her family, thus undertaking the first step toward the formalization of their relationship, but that she had refused to go live with him since, as she put it, she "could no longer stand him." According to Amelia, this often happened when one was pregnant with a boy. Nadia, another young woman from Liberdade, was pleased to introduce me to the father of her second child when he came to formally introduce himself to her mother. When I asked whether she was going to move in with him, she explained that their baby was still too young and that she needed to stay close to her mother who would know what to do were the baby ever to fall ill. She smiled when I pointed out that the father lived up the road, less than a kilometer away.

By inquiring into the whereabouts of the fathers of these and other children,[7] I was probing at an unpleasant public secret of the postwar economy. Although men's irresponsibility was certainly not fabricated, as a blanket justification for single-motherhood it hid a more complex and unsettling reality within which young women were actively and quite willfully taking advantage of postwar reconfigurations to remain uncommitted. A number of young women, like Nadia and Amelia, had come to the conclusion that they could lead more fulfilling lives by engaging in relationships with various men and had developed crafty ways to use their sexuality to their advantage. Instead of letting a pregnancy tie them down to one man, several of the young women I worked with preferred to stay put—with their own mothers—at least for the time being.[8] As I came to know some of these young women well, I also came to better understand the ramifications of their life choices.

Unlike *lobolo*, which involves transactions between men of different families, these more informal relationships allow women to channel resources and thus bypass their male kin (Hunter 2002: 112) while also keeping the flow going through the accumulation of sexual partners. But as Christian Groes-

7. A number opted for a more radical solution and performed home abortions, though sometimes with dire consequences. A number of women I worked with admitted to having themselves performed such abortions and all knew of someone that had.

8. Nearly one-fourth of all households in southern Mozambique are female-headed households, two-thirds of which are headed by widows (de Vletter 2006: 9). I would estimate that these figures are even higher in cities and certainly so in Liberdade.

Green (2014) argues, based on research among Mozambican women involved with European men, looking at these exchanges as dyadic occludes the wider kin networks within which they are embedded. A similar logic operates in Inhambane where poorer households have come to rely on the participation of their young women in the intimate economy.

The Cabral Family: Part 2

I dropped by the Cabrals' to pay them a visit like I always do whenever I am in the area. It was mid-morning and I arrived to find Julia watching television with her two-year-old grandson, the son of her youngest daughter Sandra. Julia told me that Sandra was out running errands. But she abruptly stopped mid-sentence, paused for a few seconds, and then said that since we knew each other so well (*como você é de casa*), there was no need to pretend: Sandra was still at a man's house where she had spent the night. "She's with her boss [*patrão*]!" Julia jokingly added. Sandra came home about an hour later and greeted me with a cheeky smile. Julia never really attempted to restrict her daughters' intimate affairs. Having herself had children with seven different men, she knew from experience that there was no other way for women from poor families.

We headed out for a stroll. Sandra told me that since her "boss"—not the father of her son but another man she had met a few months earlier—was from out of town, she was able to come and go as she pleased with no nosy relatives to contend with. "I live between both houses," she said, "and whenever he gives me money to buy a *rancho* [a ration of basic foodstuffs usually purchased at the beginning of the month], I divide the money and buy things twice: for his house and for mother's house." I asked her what the "boss" made of this. "Well if I already had a son when he met me," she responded, "how could he possibly complain?" Sandra then looked at her phone and said we had to hurry up a little. We took the path leading up to Mafurera market and stopped at a crossroads on the way to talk to a young man who was apparently waiting for Sandra. They exchanged a few words before he slipped her a 200 MZN note. The man appeared a little uncomfortable but we left before he had a chance to object. Sandra then filled me in: "I met this guy before meeting my boss but my boss was quicker at seducing me. When the second one finally made it clear that he was interested in me, I told him that I already had a boyfriend. But he responded: 'I don't need your boyfriend, I need you.' So now I have a system whereby I charge both! And then there's also my son's father, when he remembers to pay maintenance."

I did come across disgruntled mothers who felt that changing modes of access to money and things violated generational hierarchies. As one mother explained:

> Younger girls now own things that their mothers don't even own! That's a big lack of respect. In the past, whatever you owned, you had received from your husband. And as a child, you couldn't come home with anything, not even a sweet [without providing an explanation]. But now everything is disorganized.

Parents felt they had little authority over their children in general, and over their daughters in particular. This is how Ana, a middle-aged widowed mother of five, put it to me:

> The youth accuse us of being witches but what makes them talk this way is poverty. They insult us and kill us. They wake up in the morning and see that, again, there is nothing to eat. They ask us for food and they don't believe us when we say that we don't have any. Girls go around having sex with any men who'll give them a bit of money. They have to search outside as there is nothing at home. I made them and raised them and now instead of helping me, they come home with more children. . . .[9] In the past, young people used to respect their parents because there used to be food. Now that parents struggle to feed their children, the children don't see why they should respect their parents. Maybe it's also because we don't know how to educate them. But the problem isn't just with my kids, it's with everyone's kids [i.e., whether the household is woman-headed or not]. Kids make and break [*fazem e disfazem*].

These critiques spoke of a real discomfort but they also hid a tacit understanding that these daughters, especially the fatherless ones, had little choice other than to take part in the intimate economy. The postsocialist, postwar economy, with its redefined expectations and aspirations and its growing inequality, has put added pressures on women to find alternative sources of income rather than hope for a husband to look after them. But women also did their best to keep their options open by trying to evade being labeled as loose women. In their efforts to craft fulfilling lives, young women were actively manipulating regimes of truth and subverting preexisting categories. No one in Liberdade denied that there always were women who were willing to exchange sexual favors and who had multiple partners. Many believed, however, that mobile phones were amplifying the trend. "With mobile phones, *chular* is no longer just a game; it's a sport!" young men in a debate I organized concluded.[10] But the phone was also understood to help occlude these practices, and the blurring of boundaries between what were believed to have been discrete categories of women was described as particularly problematic by those who struggled to navigate their way among what they saw as increas-

9. At the time, Ana's two daughters were living with her as second-generation single mothers.
10. "Sport" is used as a metaphor to describe activities done in excess.

FIGURE 11. Tofo Beach on a Sunday, Inhambane, 2007. Photo by author.

ingly undifferentiated women. As Inocencio, the young man I quoted earlier, nicely put it to me: “There are two categories of girls—girls to marry and girls to play with—but the problem is that girls, these days, are very clever with their mobile phones and all, and we end up not knowing which is which!”

To Conceal Is Respect

In the chapter on love and deceit, I looked at how the phone was used in courtship as a tool of cajolery and persuasion, and at how mobile communication provided a space within which new ways of being and relating could be imagined and tried out. Mobile communication also helps bypass public scrutiny by creating an invisible realm within which intimate relationships like the ones described in the following example can be negotiated respectfully.

When I first met Bela, she was involved with a violent young man who regularly cheated on her. He was also extremely jealous and would often beat her, only to later shower her with gifts and beg for her forgiveness. Bela later left him for an educated man whom she described as romantic and respectful. By this, she meant that she had never seen him with another girl or found anything suspicious on his mobile phone. Her new boyfriend had a problem, though: he never gave her anything. As Bela put it,

> If you see me wearing nice clothes, it's quite obvious that it's not my [unemployed] father who bought them for me. But [my boyfriend] doesn't ask me where I got them. After a few months of seeing each other, he still hasn't given me anything. He never asks me where I get my perfume from. He compliments me on my hairdo but he doesn't ask me how I pay for it.

She concluded, "At least he gets upset whenever I receive a text message in his presence!" Although "a respectful man," Bela's new boyfriend was not a good provider. As she carefully monitored his activities—"I usually go through his phone when he's taking a bath," she admitted—Bela used her own phone to discreetly manage relationships with other, more generous men. Bela found that the only way she could start fulfilling some of her desires was to multiply intimate partners. However, instead of having these different men knock on her door for everyone to see, she used her phone to inconspicuously coordinate rendezvous. In doing so, Bela was able to better manipulate what would otherwise be irreconcilable ideals of respectability with economic emancipation.

It is fair to assume, as Bela herself suggested, that her boyfriend knew that she was involved with other men. Bela's boyfriend was, however, in an uncomfortable position. On the one hand, given that women are meant to remain sexually exclusive, the boyfriend would have had every right to voice disapproval. But, on the other hand, given that men are also meant to provide for women, he risked being put on the spot. In a sense, Bela's infidelity was a critique of her boyfriend's inability to live up to mainstream ideals of masculinity. Bela's boyfriend must have been torn between jealous sentiments, a frustrated sense of entitlement, and shame of his own shortcomings. If he wanted to continue seeing Bela, he had little choice but to feign ignorance, to pretend not to know, as knowing would have commanded some form of remedial action. Bela made it easier for both her boyfriend and her father to feign ignorance, to overlook the fact that she was involved with older men, by using her phone to manage these relationships as discreetly as she could. In doing so, Bela could also more easily project the image of being a good girl—a *moça de casa*—and thus preserve her and her family's respectability while also keeping her options open, in case the day came when someone asked to marry her. If, to understand the construction of sexual identity, we first need to take into account concerns about reputation (Hirsch, Wardlow, and Phinney 2012), I also insist that discretion, even if not absolute, weighs heavily in how particular individuals will be categorized. Efforts of discretion, by way of various tricks and technologies, are rewarded.

Everyone knew that their partner was likely to be intimately involved with other people and that if they set out to look for proof, they were bound to find some. "If you seek, you find" (*quem procura, encontra*), went the saying. But many preferred to give their partner the benefit of the doubt (*prefiro acreditar*). Having recently celebrated his fiftieth wedding anniversary, Emidio could not tell me with certainty whether his wife had ever cheated on him. What he knew, however, was that even if she had, she had never "shown" him ("*ela nunca me fiz ver*") that there were other men in her life. To make sure that his statement was not construed as a naïve belief in his wife's fidelity, Emidio added that women, like men, had desires and that it was therefore not unlikely. But that did not seem to matter all that much to Emidio, provided it was done respectfully. He concluded by saying, "*esconder e respeito*" (to conceal is respect).[11] Another connected saying I heard often was "one who conceals does so because he cares" (*quem esconde é porque gosta*). Indeed, as I mentioned already, a good partner was not necessarily a faithful one, but a discreet one. A good partner was a respectful one. Emidio's use of the verb *to show* highlighted the importance of visibility in regimes of truth. Until a relationship was made visible, its existence could be denied.[12]

Regimes of truth based on the visible are familiar to Western sensibilities, given that the bias toward vision as the privileged sense with which we apprehend the world dates back at least to Aristotle (Classen 1997: 402). In a short piece on love in which he reflected on the entanglement between love and knowledge, Alfred Gell ([1996] 2011: para. 5) wrote about adultery in Umeda, but in a way that could easily apply to intimate relationships in Inhambane, that "such affairs generated information, lethal information. 'Fidelity,' in the Umeda scheme of things, was not sexual fidelity as we understand it, i.e. chastity, but informational fidelity, i.e. keeping information about liaisons secret. Women were not mistrusted for dishonoring their husbands physically, but because they might betray their lovers (or husbands) verbally. Or their husbands might betray them, not by being sexually unfaithful, but by giving way to—perhaps baseless—jealousy and 'speaking' to a sorcerer. In either case it was not bodily behavior as such, but disclosure, or the possibility of disclosure, which mattered."

11. And since we were on the topic, this charming septuagenarian told me about a place he knew out of town where they had nice rooms to rent and suggested we meet there one day. He promised that no one would ever know about it!

12. When talking about a girlfriend or boyfriend, young people would qualify how serious the relationship was by specifying whether or not their partner was known to their relatives (*conhecido-a em casa*). When parents said that they did not know whether their son or daughter was in a serious relationship, it was not necessarily because they did not actually know but rather because they had not been explicitly shown the existence of such a relationship.

The newfound ability to communicate with some degree of discretion following the introduction of mobile phones has important implications in the Inhambane context where privacy is scarce, where uncertainty and social disparity have widened the gap between social ideals and actual practices, and where respect rests less on self-restraint (cf. Heald 1999) than on one's efforts at concealing anything deemed disrespectful. The phone has helped preserve this "ugly" public secret about the workings of the postwar economy, with its dependencies and interdependencies, that encourages young girls to exchange sexual favors for material gain and to accumulate intimate partners like never before. The phone, then, hides more than it reveals or, perhaps more exactly, it conceals enough to enable those who chose to, or need to, to feign ignorance.[13] The phone, in other words, is used, as part of a wider arsenal of pretense, to sustain epistemological uncertainty. As detailed earlier, mobile communication adds to other tricks of dissimulation such as using linguistic subterfuge,[14] making the most of the cover of darkness, slipping away from home and coming back undetected before sunrise, and carefully managing intimate networks, and phones are, as material objects, entangled with other technologies of concealment such as opaque black grocery bags, baggy jeans, the folds of sarongs, and tall fences.

Concealment sometimes involves keeping some in the know and others in the dark, such as when a man informs his lover that he has a wife so that the lover will know how to behave if ever the three of them cross paths. It often entails some form of collusion. On a recent field trip to Inhambane, I met up with Jhoker, who filled me in on the latest gossip. He told me who had died, who was ill, and mentioned the names of some of the young girls that had "grown up" since my last visit. He talked about the baker's daughter, a young girl who had once played with my own daughter, and about her having come of age, and said that she was even hanging out with foreign men. "*A visão daquela dama é outra* [her vision is different/unique]," he explained. "When you look at her, you can tell that she's living at a different level; in another world." Although the baker was most certainly aware that his daughter now possessed things that only men could have offered her, he was most probably unaware of the scale of her endeavors. As Jhoker put it, "You have to be living

13. Speaking about Mehinaku secrecy in the Amazon, Gregor (1977) similarly explains that "techniques of concealing sexual relationships are at best holding operations; their aim is not to obscure a sexual liaison totally, but merely to make it possible to ignore it" (140).

14. A lovely example of how this operates can be found in Bellman's (1984) ethnography of the Poro ritual in which he examines the workings of this dialectical relationship and the role metaphors or "deep talk" play in the process of concealment.

in this same world of parties to know these things." We also talked about another young woman who had only just turned thirteen and was working at one of the local *baracas*. The step-father and the brother she was living with at the time were both unemployed, and they were struggling to get by. Jhoker, who knew the family well, said with a disconcerted yet understanding tone: "They've ruined the girl [*estragaram a menina*]! Now that she works in a bar anyone can have her. She is for the taking. Even I can come around later and have her in exchange for a bit of change." In this case, the family was in such a dire situation that it had forfeited its daughter/sister's respectability for a few hundred meticais. This case was also an example of a public secret, but one that was particularly disquieting for its lack of subtlety.

There are, in short, different levels of ignorance: some do not know, some pretend not to know, some choose not to know, some know but without knowing the details, and some know, or at least think they know. The discretion afforded by mobile phone communication is arguably marred, if only because of gossip and hearsay and, as such, the phone is an imperfect tool of pretense. In fact, as discussed in chapter 4, the phone often does quite the opposite by revealing through intercepted phone calls and compromising text messages information that was meant to remain secret. But the phone nonetheless conceals enough to allow those who choose to, or need to, to pretend not to know.

Secrecy participates in the reproduction of power and has often been understood as helping maintain the status quo (G. Jones 2014: 55). In his work on public secrets in public health, for example, Geissler (2013) states that public secrets are "constitutive of social order through a double bond with power: making domination unspoken, silencing critique and resistance, and exacerbating power differentials, since the force of making violence unknowable exceed that of the violent act itself" (15). In this public health context, unknowing ultimately facilitates research (Geissler 2013), just like unknowing in Liberdade facilitates intimacy. Regarding the "ugly" public secret about young women's involvement in the sexual economy, however, the implications are potentially far more subversive. That is to say that the part ignorance plays in sustaining configurations of power is at least ambiguous rather than clearly to the advantages of a patriarchal authority that finds itself, in practice, increasingly undermined. As such, the discretion granted by mobile communication is simultaneously transformative and conservative.

What happens, then, to the anthropological project of disclosure in a social world embroiled in the politics of pretense? Geissler (2013) leaves the question open when he suggests that "if everybody involved in a collaboration can tell truths from untruths and engage both registers simultaneously, anthropologists add little by 'telling the truth' and even less by relativist denial of such a distinction" (30). The ethnographer's task, then, is not so much to expose these secrets as to unveil their power (Shryock 2004: 19).

The spread of mobile phones has allowed individuals to build on the value of discretion by creating new spaces of intimacy. Mobile phone communication therefore not only helps redress broken-down boundaries, it also enables the erection of new ones, however porous, that are better tailored to the postwar economy in a place where most people have much to hide and where what counts as true is what is known—what one sees and is made to see. Mobile phone communication mitigates dissonance not by tackling its underlying roots but rather by covering them up. If we understand uncertainty as Whyte (2009) defines it, as "a lack of absolute knowledge" (213), then uncertainty in Inhambane is not always something to transcend or overcome. Liberdade youth recognize that in order to live—rather than merely survive—they need to "know what not to know" (Taussig 1999: 223). The crafting of fulfilling lives rests on a careful balance between visibility and invisibility whereby ignorance and opacity are sometimes as sought-after as knowledge and clarity. By this I mean that opacity is not simply something to foist onto others but also something that one might impose on oneself. Or, to put it differently, unlike uncertainty, which is valued precisely for its open-endedness, certainty is a hope killer—it forecloses possibilities.

CONCLUSION

Mobile Phones and the Demands of Intimacy

Estamos juntos (We are together)
MCEL SLOGAN

Anthropology is, in many ways, all about intimacy. The stuff of social relations, intimacy is not only what we study, even if often only implicitly, but also the very cornerstone of ethnography. Intimacy is, however, a rarely problematized "slippery" concept often wrongly presumed to be a "domain of harmony" (Geschiere 2013: xix, 28). As Peter Geschiere (2013) reminds us, in a recent effort to revisit his earlier writings on witchcraft as "the dark side of kinship" (Geschiere 1997), harm and suspicion are often most manifest at a very intimate level. It follows that trust, as Simmel suggested some time ago, can never be a certainty, rather, trust, in Geschiere's (2013) words, "is always precarious and situational (never 'ontological')" (xiii). Harm and deceit, as Mozambicans know very well, often comes from intimates—it is children who accuse their mothers of being witches, lovers who "put men in bottles," and friends who seduce each other's partners—and suspicion and distrust permeate the social fabric. Intimate networks are unreliable, like the country's mobile networks, and potentially damaging. The youth introduced in this book not only are reluctant to trust anyone for fear of being taken advantage of—of being robbed, duped, or bewitched—but also avoid being known by carefully juggling visibility and invisibility. Yet, despite pervasive suspicion, despite the prospects of imminent harm, people still develop and maintain intimate relationships with relatives, neighbors, partners, and lovers (Geschiere 2013). Indeed, the Africanist literature clearly shows that it is by engaging in "thick sociality" (Cooper and Pratten 2015)—AbdouMaliq Simone (2004) proposes the notion of "people as infrastructure"—that individuals navigate everyday uncertainty. In this book, I have examined how young Mozambicans were using mobile communication to address the demands of intimacy: the entanglement of obligations, necessities, suspicions, fears, desires, pains, and pleasures that make up intimacy. In this conclusion, I propose to expand on the demands of intimacy as a notion applicable beyond the Inhambane context, before coming back to the discussion on the phone's transformative potential with which I began.

The phone brings people together, as Mozambique's network provider, mCel, succinctly puts it. This is also clear in Portuguese as the term *ligar* aptly means both to connect and to phone. The connection is, however, often achieved through some form of disconection: the phone helps disseminate *and* dissimulate information. As a relational tool, the phone also encapsulates the tensions between the desire (and pressures) to redistribute and the pull of self-indulgence. Inspired by Mirjam de Bruijn and Rijk van Dijk (2012), who encourage a shift of focus away from connective technologies—the bridge, the phone, the road—toward connectivity, I have also been interested in exploring the new intimacies that mobile communication encourages and sheds light on. In doing so, it became clear that connectivity could, like intimacy more broadly, be both productive and destructive. Togetherness often entails some form of disconnection, somewhere—something Mozambicans hint at when they speak of the phone as a necessary evil.

In his work in Sierra Leone, Michael Jackson (1998) nicely captures the "conundrum of intimacy" (Geschiere 2013: 89) by showing how, although the Kuranko are preoccupied with marking perimeters, especially of their bodies and houses, with protective medicines to evade supernatural threats, they also recognize, as Jackson notes, that the "passage across these very borders is vital to life and livelihood" (43–50). Building on Husserl's notion of intersubjectivity as well as on Hannah Arendt's writings, Jackson (2013) writes: "One becomes aware of oneself through relations with others. A sense of one's own uniqueness and autonomy emerges, therefore, not from within oneself but from within contexts of intersubjective relations" (18). This relationality is, in short, fraught with difficulties and charged with the tension between the desire for individuation and autonomy, and the desire to connect with others (Jackson 2013: 6). For Jackson, it is by juggling this tension that we forge existentially viable lives. Negotiating the demands of intimacy involves not only seeing the other as a potential provider of resources, opportunities, and pleasure, but also as fundamental to recognition. It is our quest for ontological security that, for Jackson (1998), is at the heart of our common humanity, the desire "to experience the world as a subject and not solely as a contingent predicate" (71, 21). I think this is also what Elizabeth Povinelli (2006) has in mind when she examines the tensions between the "autological subject" geared toward self-making and the "genealogical society" with its constraints and pressure. It is certainly what Liberdade youth have in mind when they talk about living rather than merely surviving, and what I aim to capture when I talk about their efforts to negotiate the demands of intimacy. If the phone has proven so irresistible, it is in large part owing to how it allows people to juggle, however imperfectly, the demands of intimacy.

For Simmel ([1900] 1978), these tensions are resolved through trust. He writes: "Without the general trust that people have in each other, society itself would disintegrate, for very few relationships are based entirely upon what is known with certainty about another person, and very few relationships would endure if trust were not as strong as, or stronger than, rational proof or personal observation" (191). I showed how Liberdade youth were, for their part, concerned with erecting boundaries designed to offer protection from more mundane invasions, namely to elude the prying eye of others, how they envisaged life as an individual battle, and how they were weary of ever letting their guard down. The leap of faith that Simmel had in mind when discussing trust in *The Philosophy of Money* was, however, in such contradiction with the basic premise of *visão* that when Liberdade youth did seemingly trust each other, it was through willful blindness rather than through a suspension of doubt. Indeed, it was by maintaining open-endedness, by sustaining uncertainty, that intimacy was possible.

With the spread of neoliberal values of transparency (Mazzarella 2006; Trottier 2012) in a media-saturated world in which personal privacy and collective security are increasingly constructed by governments as being at odds with each other, an exploration into mobile communication and discretion is particularly pressing. Never have we been able to expose more publicly the most intimate details of our private lives. Yet any attempt to infringe on our privacy, any crack at going through our personal exchanges, whether by a jealous girlfriend or the National Security Agency, provokes a visceral reaction, even among those who may not have much to hide. It is a matter of principle. The clash between the FBI and Apple over access to the iPhone belonging to one of the 2016 San Bernardino shooters brought the ongoing debate around security and privacy to a new level. Asked to build a "backdoor to the iPhone" that would allow access to encrypted information that Apple itself has no access to, the technology giant has upheld values of individual freedom.[1] In a letter to customers, Tim Cook, Apple's CEO, described the implications of

1. Then again, when hackers leaked the profiles of thirty million users from across the world of the "infidelity website" Ashley Madison, specially designed to facilitate affairs between married people, there was little compassion for the victims whose privacy had been infringed upon (Lamont 2016).

the government's demands as "chilling"[2] and appealed to the public's fears of surveillance to support the company's refusal to develop software that would help unlock the iPhone. The Mozambican government is also attempting to exert more control over mobile communication. In the wake of the 2010 strikes when protestors took to the streets in Maputo and other larger cities in Mozambique to complain about rising food and fuel prices, the government passed a law requiring all pay-as-you-go clients to register their SIM cards (Bertelsen 2014). The deadline for doing so was extended a number of times, but in March 2016 a million people who had failed to register were disconnected.[3] These are unfolding stories fought as legal battles but that speak of a global climate of fear and suspicion.

In the introduction, I toyed with the idea that mobile phones might save Africa to encourage a serious reflection on the phone's transformative potential. Throughout the book, I showed how mobile phone practices in Inhambane were shaped by the local political economy of display and disguise, and situated the phone within a wider arsenal of pretense. Like Horst and Miller (2006) in their study of phone practices in Jamaica, and Daniel Jordan Smith (2006) in his analysis of the phone's impact in Nigeria, I highlighted how the phone had, as Smith puts it, "accentuated already prevalent cultural dynamics" (498–99). Fischer (1992) had reached a similar conclusion regarding the introduction of landlines into American households. Fischer concluded that "the telephone did not radically alter American ways of life; rather, Americans used it to more vigorously pursue their characteristic ways of life" (5). This, however, is not to say that important social reconfigurations are not enhanced or even triggered by the adoption of new communication technologies. The domain of what is possible and feasible has, indeed, been undeniably altered with the spread of telecommunication. I also showed how mobile communication was giving rise to new intimacies by opening up new intimate spaces within which new ways of being and relating could be explored. It is this transformative potential that inspires awe and triggers moral panic. Indeed, while individuals may very well use the phone to be "more like themselves" (Sahlins 1993: 17), they also use new technologies in their attempts to transcend the constraints of being "like themselves" (cf. West 2005).

We were, it seems to me, probably too quick at dismissing ICT4D enthusiasm as disconnected from empirical reality. While my findings would un-

2. The letter is available on the Apple website, http://www.apple.com/customer-letter/, accessed March 5, 2016.

3. According to the Mozambique News Agency website, http://www.poptel.org.uk/mozambique-news/newsletter/aim525.html#story12, accessed March 9, 2016.

doubtedly make most proponents of ICT4D grimace—the intimate economy, as described above, is far removed from ideas of ICT-induced development based on the circulation of useful information—I would argue that they help qualify rather than all together challenge the ICT4D framework. The spread of mobile phones has facilitated access to information and improved people's quality of life in a tangible fashion by helping lubricate the circulation of money and other resources, as well as in a more subtle manner, by providing individuals with a certain degree of authorship and control over their lives, albeit in gender-specific and contested ways, even if longer term implication may very well be rather disastrous.[4]

If, then, we accept that changes in modes of access to information have a transformative potential, we must bear in mind that what constitutes "useful information" varies considerably depending on the sociohistorical context, not to mention on where one stands in a deceitful relationship. If the phone's entanglement in intimate affairs seems mundane and inconsequential in comparison to questions of access to market prices or medical diagnosis, for example, this ethnographic account suggests that, by helping subvert gender expectations, the role mobile phones plays in everyday sociality might in fact have much broader transformative implications, the ramifications of which are only starting to emerge.

Telecommunication helps lubricate, strengthen, and expand redistribution networks. In doing so, it contributes to the leveling of growing disparity by facilitating the circulation of money and things between "those who have" (*os que tem*) and "those who ask" (*os que pedem*). By making these exchanges more efficient, mobile communication does, however, also render social networks potentially more extractive. Various authors have shown how kinship obligations inhibit development by channeling resources that end up consumed rather than invested to generate wealth (Hanlon 2007: 12–13). Owing to the phone's role in enhancing and transforming redistribution networks, the

4. When considering, for example, the role mobile communication plays in increasing invisibility and facilitating the accumulation of sexual partners, the long-term impacts on the transmission of HIV/AIDS and other sexually transmitted infections may be tragic (cf. Hirsch et al. 2010). This is only a hypothesis, as I have not conducted research on HIV transmission. The link between transactional sex and the transmission of HIV/AIDS is discussed by many, including Hunter (2002). And although sub-Saharan Africa may no longer be excluded from the information age (cf. Castells 2000), the worldwide uptake of mobile phones nonetheless indirectly sustains older cleavages. The rising global demand for minerals such as coltan—an essential component in mobile phone manufacturing—has been feeding a bloody conflict in Congo where 80 percent of the world's reserve is believed to lie (Mantz 2008; J. Smith 2011), while other parts of the continent have become dumping grounds for e-waste (Burrell 2012).

implications are therefore potentially compounded. In the process, new forms of dependency and exclusion are starting to emerge. It is in fact ironic that such a potent symbol of the twenty-first century can actually distance people from capitalistic ideals. The widespread adoption of mobile phones in contexts of underdevelopment also raises important questions about the allocation of scarce resources. The proportion that low-income households spend on mobile communication is often staggering. In assessing these practices and their economic implications, we need to remember that individuals operate in complex, often contradictory ways and determine their priorities in socioculturally inflected ways. As Douglas and Ney (1998) figuratively put it: "Not nature but culture defines what a full belly is, how full it should be, and what is needed to fill it" (49). The different ways in which young people in Mozambique use their phones in their attempts to lead fulfilling lives certainly highlights how ill-founded it would be to distinguish necessities from luxuries. Still, if poorer people are pretty good at getting others to subsidize their phone calls, and if they also often use their phones to ask for money and things, it is difficult to determine how the economic costs and benefits tally over a period of time. Future studies interested in assessing the impacts of mobile phones should therefore look beyond the conventional areas of socioeconomic development (business, education, healthcare, and governance).

The ethnographic investigation of mobile phone practices provides a useful way of understanding what it might entail to be a young adult living in postwar, postsocialist Mozambique. Mobile phones increasingly feature in our ethnographies, even when these are not specifically focused on phone use. As anthropologists, we do, however, also need to start reflecting on the ways in which widespread access to mobile communication impacts ethnography, and anthropological research more broadly.

As I write these last pages, I have in mind all that has changed in Inhambane since I first set foot in the land of good people over a decade ago. Where cars were few and far between, there is now the occasional traffic jam. Brick houses under construction dot the peri-urban landscape where there used to be only palm trees and the odd cashew tree. The city is expanding rapidly. There are now daily flights to and from Johannesburg, and there is a new bus service that links Inhambane to Tofo Beach. Several of my friends have moved away, others have passed away, and those who remain still "travel while sitting down" (Archambault 2012b). Yet the city and its residents are, overall, much better connected. Today, young people also have access to superior phones, many of which are Internet enabled, and are therefore using mobile communication in new and exciting ways. Still, in their engagement with social media, their preoccupations echo earlier efforts to juggle visibility and invisibility.

Ideals of masculinity, like those of femininity, come with a specific set of expectations and entitlements. But each also has its distinct temporalities. While young women and their families may be reaping the benefits of their involvement in the intimate economy, they may struggle to carry on doing so as they get older. Jennifer Cole (2010) has shown, for example, how women in Madagascar often turned to religion as they aged after spending their youth participating in the sexual economy.

Within the cohort of young people that I have been following since 2006, the majority of those with a secondary school diploma have since secured a job. Inocencio teaches English at a local secondary school, Jhoker works at the meteorology station, Fifi was promoted from printing clerk at the university to personal assistant of the director, and Kenneth is helping start up a branch of the Catholic University in Maputo, to name only a few examples. A number have even started building their own houses, a key marker of adulthood throughout sub-Saharan Africa that, until recently, was, for most, at best a deferred dream. As Bush, who has a knack for putting things clearly, told me during a follow-up field trip: "The last time you were here, we were all busy with mobile phones but now it's building that's in fashion!" Testifying to Africa's rapid urbanization, these trends also add nuance to understandings of the struggles of adulthood. I can only hope that Frelimo and Renamo will reach an agreement and put an end to the fighting in the center of the country. The word on the street is that Mozambicans do not want to go back to war. Certainly in Inhambane, perhaps for the first time in a long time, a growing number of young people actually have something to lose.

Over the years, many people in Inhambane have told me that they had plans to one day write a book that would recount their unique life story. *Minha Vida* (My life) was the proposed title of many of these imagined books. These autobiographical projects were meant to showcase one's originality and distinction, rather than one's hardships, their own version of praise poetry, which, as described by Karin Barber (2007), is "the strongest affirmation of personal distinctiveness and individual agency: it is about making a mark, being outstanding, doing extraordinary and memorable deeds" (112). In the meantime, here is my version. I hope to have conveyed the very articulate and often colorful ways in which my young interlocutors engage with and see the world, with all its unsettling but also tantalizing uncertainty.

If this book achieves only one thing, I hope it gives a sense of what it might feel like to be a young person growing up in Mozambique today. Although they often described life as a battle, young people in Liberdade insisted that they wanted to live and not merely survive, and although everyday life was often tumultuous, they expended a lot of energy in making it seem as though everything was running smoothly, as though they were effortlessly cruising through uncertainty. To embark on this cruise, to live rather than merely survive, owning a mobile phone had become a necessity.

References

Abbink, Jon. 2004. "Being Young in Africa: The Politics of Despair and Renewal." In *Vanguard or Vandals: Youth, Politics and Conflict in Africa*, edited by J. Abbink and I. V. Kessel, 1–34. Leiden: Brill.

Abrahams, Roger D. 1970. "A Performance-Centered Approach to Gossip." *Man NS* 5(2): 290–301.

Abrahams, Roger D. 1983. *The Man-of-Words in the West Indies: Performance and the Emergence of Creole Culture*. Baltimore: Johns Hopkins University Press.

Adams, Mary. 2009. "Playful Places, Serious Times: Young Women Migrants from a Peri-Urban Settlement, Zimbabwe." *Journal of the Royal Anthropological Institute* 15(4): 797–818.

Agadjanian, Victor. 2005. "Men Doing 'Women's Work': Masculinity and Gender Relations among Street Vendors in Maputo, Mozambique." In *African Masculinities: Men in Africa from the Late Nineteenth Century to the Present*, edited by L. Ouzgane and R. Morrell, 257–69. New York: Palgrave Macmillan.

Ahearn, Laura M. 2001. *Invitations to Love: Literacy, Love Letters, and Social Change in Nepal*. Ann Arbor: University of Michigan Press.

Aker, Jenny C., Paul Collier, and Pedro C. Vicente. 2011. "Is Information Power? Using Mobile Phones During an Election in Mozambique." Accessed January 21, 2013. http://www.pedrovicente.org/cell.pdf.

Akyeampong, Emmanuel Kwaku. 1996. *Drink, Power and Cultural Change: A Social History of Alcohol in Ghana, c. 1800 to Recent Times*. Portsmouth, NH: Heinemann.

Alexander, Jocelyn. 1997. "The Local State in Post-War Mozambique: Political Practice and Ideas About Authority." *Africa* 67: 1–26.

Alpers, Edward A. 2009. *East Africa and the Indian Ocean*. Princeton: Markus Wiener Publishers.

Amit, Vered, and Noel Dyck. 2012. "Pursuing Respectable Adulthood." In *Young Men in Uncertain Times*, edited by V. Amit and N. Dyck, 1–31. New York: Berghahn Books.

Amit-Talai, Vered, and Helena Wulff, eds. 1995. *Youth Cultures: A Cross-Cultural Perspective*. London: Routledge.

Andersen, Barbara. 2013. "Tricks, Lies, and Mobile Phones: 'Phone Friend' Stories in Papua New Guines." *Culture, Theory and Critique* 54(3): 318–34.

Anderson, Benedict. 1983. *Imagined Communities: Reflections on the Origin and Spread of Nationalism*. London: Verso.

Appadurai, Arjun. 1990. "Disjuncture and Difference in the Global Cultural Economy." *Public Culture* 2(2): 1–24.

Archambault, Julie Soleil. 2009. "Being Cool or Being Good: Researching Mobile Phones in Southern Mozambique." *Anthropology Matters* 11(2): 1–9.

Archambault, Julie Soleil. 2011. "Breaking Up 'Because of the Phone' and the Transformative Powers of Information in Southern Mozambique." *New Media & Society* 13(3): 444–56.

Archambault, Julie Soleil. 2012a. "Mobile Phones and the 'Commercialisation' of Relationships: Expressions of Masculinity in Southern Mozambique." In *Super Girls, Gangstas, Freeters, and Xenomaniacs: Gender and Modernity in Global Youth Cultures*, edited by K. Brison and S. Dewey, 47–71. Syracuse: Syracuse University Press.

Archambault, Julie Soleil. 2012b. "'Travelling While Sitting Down': Mobile Phones, Mobility and the Communication Landscape in Inhambane, Mozambique." *Africa* 82(3): 392–411.

Archambault, Julie Soleil. 2013. "Cruising through Uncertainty: Cell Phones and the Politics of Display and Disguise in Inhambane, Mozambique." *American Ethnologist* 40(1): 88–101.

Archambault, Julie Soleil. 2014. "Rhythms of Uncertainty and the Pleasures of Anticipation." In *Ethnographies of Uncertainty in Africa*, edited by E. Cooper and D. Pratten, 129–48. New York: Palgrave Macmillan.

Archambault, Julie Soleil. 2016. "Taking Love Seriously in Human-Plant Relations in Mozambique: Towards an Anthropology of Affective Encounters." *Cultural Anthropology* 31(2): 244–71.

Arndt, C., R. C. James, and K. R. Simler. 2006. "Has Economic Growth in Mozambique Been Pro-Poor?" *Journal of Southern African Economies* 15(4): 571–602.

Arnfred, Signe. 2001. *Family Forms and Gender Policy in Revolutionary Mozambique (1975–1985)*. Bordeaux: Centre d'études d'Afrique noire.

Arnfred, Signe. 2011. *Sexuality and Gender Politics in Mozambique: Rethinking Gender in Africa*. Woodbridge, UK: James Currey.

Bagnol, B., and E. Chamo. 2003. *'Titios' E 'Catorzinhas': Pesquisa Exploratória Sobre 'Sugar Daddies' Na Zambézia (Quelimane E Pebane)*. DFID/PMG Mozambique. Accessed March 6, 2012. http://www.wlsa.org.mz/wpcontent/uploads/2014/11/SugarDaddies.pdf.

Barber, Karin. 2007. *The Anthropology of Texts, Persons and Publics*. Cambridge: Cambridge University Press.

Barker, Joshua. 2008. "Playing with Publics: Technology, Talk and Sociability in Indonesia." *Language and Communication* 28: 127–42.

Barnes, J. A. 1994. *A Pack of Lies: Towards a Sociology of Lying*. Cambridge: Cambridge University Press.

Bastian, Misty L. 2000. "Young Converts: Christian Missions, Gender and Youth in Onitsha, Nigeria, 1880–1929." *Anthropological Quarterly* 73(3): 145–58.

Batson-Savage, T. 2007. "'Hol' Awn Mek a Answer Mi Cellular': Sex, Sexuality and the Cellular Phone in Urban Jamaica." *Continuum: Journal of Media and Cultural Studies* 21(2): 239–51.

Bayart, Jean-François. 1989. *L'état En Afrique: La Politique Du Ventre*. Paris: Fayart.

Bellman, Beryl Larry. 1984. *The Language of Secrecy: Symbols and Metaphors in Poro Ritual*. New Brunswick, NJ: Rutgers University Press.

Berker, Thomas, Maren Hartmann, Yves Punie, and Katie J. Ward. 2006. "Introduction." In *Domestication of Media and Technology*, edited by T. Berker, M. Hartmann, Y. Punie, and K. J. Ward, 1–17. Glasgow: Bell and Bain Ltd.

Bernstein, Elizabeth. 2007. *Temporarily Yours: Intimacy, Authenticity, and the Commerce of Sex*. Chicago: University of Chicago Press.

Bertelsen, Bjorn Enge. 2014. "Effervescense and Ephemerality: Popular Urban Uprisings in Mozambique." *Ethnos*: 1–28.

Berthomé, François, Julien Bonhomme, and Grégory Delaplace. 2012. "Preface: Cultivating Uncertainty." *HAU: Journal of Ethnographic Theory* 2(2): 129–37.

Bledsoe, Caroline, and Gilles Pison. 1994. "Introduction." In *Nuptiality in Sub-Saharan Africa*, edited by C. Bledsoe and G. Pison, 1–22. Oxford: Clarendon Press.

Bloch, Maurice. 1995. "People into Places: Zafimaniry Concepts of Clarity." In *The Anthropology of Landscape. Perspectives on Place and Space*, edited by E. Hirsch and M. O'Hanlon, 63–77. Oxford: Oxford University Press.

Boellstorff, Tom. 2008. *Coming of Age in Second Life*. Princeton: Princeton University Press.

Bourgois, Philippe. 2002. "Understanding Inner-City Poverty: Resistance and Self-Destruction under U.S. Apartheid." In *Exotic No More: Anthropology on the Front Lines*, edited by J. MacClancey, 15–32. Chicago: University of Chicago Press.

Bourgois, Philippe. 1996. *In Search of Respect: Selling Crack in El Barrio*. Cambridge: Cambridge University Press.

Breckenridge, Keith. 2006. "Reasons for Writing: African Working-Class Letter-Writing in Early-Twentieth-Century South Africa." In *Africa's Hidden Histories: Everyday Literacy and Making the Self*, edited by K. Barber, 143–54. Bloomington: Indiana University Press.

Breslauer, D. N., R. N. Maamari, N. A. Switz, W. A. Lam, and D. A. Fletcher. 2009. "Mobile Phone Based Clinical Microscopy for Global Health Applications." *PLoS ONE* 4(7): e6320.

Brison, Karen, and Susan Dewey, eds. 2012. *Super Girls, Gangstas, Freeters, and Xenomaniacs: Gender and Modernity in Global Youth Cultures*. Syracuse: Syracuse University Press.

Brooks, Andrew. 2012. *Riches from Rags or Persistent Poverty? Inequality in the Transnational Second-Hand Clothing Trade in Mozambique*. PhD diss., Royal Holloway.

Bucholtz, Mary. 2002. "Youth and Cultural Practice." *Annual Review of Anthropology* 31: 525–52.

Buckingham, David, ed. 2008. *Youth, Identity, and Digital Media*. Digital Media and Learning. Cambridge, MA: MIT Press.

Bull, Michael. 2004. "Sound Connections: An Aural Epistemology of Proximity and Distance in Urban Culture." *Environment and Planning* 22: 103–16.

Burke, Timothy. 1996. *Lifebuoy Men, Lux Women: Commodification, Consumption and Cleanliness in Modern Zimbabwe*. London: Leicester University Press.

Burr, Lars. 2010. "Xiconhoca: Mozambique's Ubiquitous Post-Independence Traitor." In *Traitors: Suspicion, Intimacy and the Ethics of State-Building*, edited by S. Thiranagama and T. Kelly, 24–47. Philadelphia: University of Pennsylvania Press.

Burrell, Jenna. 2008. "Problematic Empowerment: West African Internet Scams as Strategic Misrepresentation." *Information Technologies and International Development* 4(4): 15–30.

Burrell, Jenna. 2009. "Could Connectivity Replace Mobility? An Analysis of Internet Café Use Patterns in Accra, Ghana." In *Mobile Phones: The New Talking Drums of Africa*, edited by M. de Bruijn, F. Nyamnjoh, and I. Brinkman, 151–69. Leiden: Langaa & African Studies Centre.

Burrell, Jenna. 2010. "User Agency in the Middle-Range: Rumors and the Reinvention of the Internet in Accra, Ghana." *Science, Technology, and Human Values* 35(5): 139–59.

Burrell, Jenna. 2012. *Invisible Users: Youth in the Internet Cafes of Urban Ghana*. Cambridge, MA: MIT Press.

Butler, Rhette. 2005. "Cell Phones May Help 'Save' Africa." Accessed March 4, 2012. http://news mongabay.com/2005/0712-rhet_butler.html.

Cahen, Michel. 2000. "L'état Nouveau Et La Diversification Religieuse Au Mozambique, 1930–1974." *Cahiers d'Études africaines* 158(XL-2): 309–49.

Cahen, Michel. 2004. *Os Outros. Um Historiador Em Moçambique, 1994*. Basel: P. Schlettwein Publishing.

Caldeira, Teresa P. R. 2000. *City of Walls: Crime, Segregation and Citizenship in São Paulo*. Berkeley: University of California Press.

Cammack, Diana. 1987. "The 'Human Face' of Destabilization: The War in Mozambique." *Review of African Political Economy* 14(40): 65–75.

Candea, Matei. 2011. "Endo/Exo." *Common Knowledge* 17(1): 156–60.

Carey, Matthew. 2012. "'The Rules' in Morocco? Pragmatic Approaches to Flirtation and Lying." *HAU: Journal of Ethnographic Theory* 2(2): 188–204.

Carmona, F. 2007. "Guebuza Na Vodacom." *Savana* XIII: 1, 3. Maputo.

Castel-Branco, Carlos Nuno. 2009. "Economia Política Da Fiscalidade E a Indústria Extractiva." Paper presented at Second Conference of the Institute of Social and Economic Studies, Maputo, Mozambique.

Castells, Manuel. 2000. *The Rise of the Network Society*. The Information Age: Economy, Society and Culture. Oxford: Wiley-Blackwell.

Chingono, Mark F. 1996. *The State, Violence and Development: The Political Economy of War in Mozambique, 1975–1992*. Aldershot, UK: Avebury.

Classen, Constance. 1997. "Foundations for an Anthropology of the Senses." *International Social Sciences Journal* 49(153): 401–12.

CNN. 2008. *Cell Phones in Africa*. Cable News Network. USA. Accessed March 4, 2012. http://transcripts.cnn.com/TRANSCRIPTS/0803/08/i_if.01.html.

Cole, Jennifer. 2010. *Sex and Salvation: Imagining the Future in Madagascar*. Chicago: University of Chicago Press.

Cole, Jennifer. 2014. "The Télèphone Malgache: Transnational Gossip and Social Transformation among Malagasy Marriage Migrants in France." *American Ethnologist* 41(2): 276–89.

Comaroff, Jean, and John L. Comaroff. 2004. "Notes on Afromodernity and the Neo World Order: An Afterword." In *Producing African Futures: Ritual and Reproduction in a Neoliberal Age*, edited by B. Weiss, 329–48. Leiden: Brill.

Comaroff, Jean, and John L. Comaroff. 2012. *Theory from the South*. Boulder: Paradigm.

Constable, Nicole. 2009. "The Commodification of Intimacy: Marriage, Sex, and Reproductive Labor." *Annual Review of Anthropology* 38: 49–64.

Cooper, Elizabeth, and David Pratten. 2015. "Ethnographies of Uncertainty in Africa: An Introduction." In *Ethnographies of Uncertainty in Africa*, edited by E. Cooper and D. Pratten, 1–16. New York: Palgrave Macmillan.

Cooper, Frederick. 2002. *Africa since 1940: The Past of the Present*. Cambridge: Cambridge University Press.

Corbett, Sara. 2008. "Can the Cellphone Help End Global Poverty?" *New York Times*, April 13. http://www.nytimes.com/2008/04/13/magazine/13anthropology-t.html?_r=0.

Cornwall, Andrea. 2002. "Spending Power: Love, Money, and the Reconfiguration of Gender Relations in Ado-Odo, Southwestern Nigeria." *American Ethnologist* 29(4): 963–80.

Cornwall, Andrea. 2003. "To Be a Man Is More Than a Day's Work: Shifting Ideals of Masculinity in Ado-Odo, Southwestern Nigeria." In *Men and Masculinities in Modern Africa*, edited by L. A. Lindsay and S. F. Miescher, 230–48. Portsmouth, NH: Heinemann.

Crapanzano, Vincent. 2004. *Imaginative Horizons: An Essay in Literary-Philosophical Anthropology*. Chicago: University of Chicago Press.

Crentsil, Perpetual. 2013. "From Personal to Public Use: Mobile Telephony as Potential Mass Educational Media in HIV/AIDS Strategies in Ghana." *Journal of the Finnish Anthropological Society* 38(1): 83–101.

Cruise O'Brien, Donald B. 1996. "A Lost Generation: Youth Identity and State Decay in West Africa." In *Postcolonial Identities in Africa*, edited by R. Werbner and T. Ranger. London: Zed Books.

Cruz e Silva, Teresa. 1998. "Identity and Political Consciousness in Southern Mozambique, 1930–1974: Two Presbyterian Biographies Contextualised." *Journal of Southern African Studies* 24(1), Special Issue on Mozambique: 223–36.

Cruz e Silva, Teresa. 2001. *Protestant Churches and the Formation of Political Consciousness in Southern Mozambique (1930–1974)*. Basel: P. Schlettwein Publishing.

Davidson, Joanna. 2010. "Cultivating Knowledge: Development, Dissemblance, and Discursive Contradictions among the Diola of Guinea-Bissau." *American Ethnologist* 37(2): 212–26.

De Boeck, Filip. 2004. "On Being *Shege* in Kinshasa: Children, the Occult and the Street." In *Reinventing Order in the Congo: How People Respond to State Failure in Kinshasa*, edited by T. Trefon, 155–73. London: Zed Books.

De Boeck, Filip, and Alcinda Honwana. 2005. "Introduction: Children and Youth in Africa: Agency, Identity and Place." In *Makers and Breakers: Children & Youth in Postcolonial Africa*, edited by A. Honwana and F. De Boeck, 1–18. Oxford: James Currey.

de Bruijn, Mirjam, Francis Nyamnjoh, and Inge Brinkman. 2009. "Introduction." In *Mobile Phones: The New Talking Drums of Africa*, edited by M. de Bruijn, F. Nyamnjoh, and I. Brinkman, 11–22. Leiden: Langaa & African Studies Centre.

de Bruijn, Mirjam, and Rijk van Dijk. 2012. "Connectivity and the Postglobal Moment: (Dis) Connections and Social Change in Africa." In *The Social Life of Connectivity in Africa*, edited by M. de Bruijn and R. van Dijk, 1–20. New York: Palgrave Macmillan.

de la Cadena, Marisol. 2010. "Indigenous Cosmopolitics in the Andes: Conceptual Reflections Beyond 'Politics.'" *Cultural Anthropology* 25(2): 334–70.

de Sola Pool, Ethiel, ed. 1977. *The Social Impact of the Telephone*. Cambridge, MA: MIT Press.

de Vletter, Fion. 2006. *Migration and Development in Mozambique: Poverty, Inequality and Survival*. Cape Town: Idasa.

DeWalt, Kathleen M., and Billie R. DeWalt. 2002. *Participant Observation: A Guide for Fieldworkers*. Walnut Creek, CA: Altamira.

Dewey, John. 1929. *The Quest for Certainty*. New York: Minton, Balch & Company.

Dinerman, Alice. 2006. *Revolution, Counter-Revolution and Revisionism in Postcolonial Africa: The Case of Mozambique, 1975–1994*. London: Routledge.

Di Nunzio, Marco. 2015. "Embracing Uncertainty: Young People on the Move in Addis Ababa's Inner City." In *Ethnographies of Uncertainty in Africa*, edited by E. Cooper and D. Pratten, 149–72. New York: Palgrave Macmillan.

Donner, Jonathan. 2008. "Research Approaches to Mobile Use in the Developing World: A Review of the Literature." *Information Society* 24: 140–59.

Donner, Jonathan, and Patricia Mechael, eds. 2013. *mHealth in Practice*. London: Bloomsbury.

Douglas, Mary, and Steven Ney. 1998. *Missing Persons: A Critique of Personhood in the Social Sciences*. Berkeley: University of California Press.

Douglas, Mary, and Baron Isherwood. 1979. *The World of Goods: Towards an Anthropology of Consumption*. London: Routledge.

Duncombe, Richard. 2012. "An Evidence-Based Framework for Assessing the Potential of Mobile Finance in Sub-Saharan Africa." *Journal of Modern African Studies* 50: 369–95.

Durham, Deborah. 2004. "Disappearing Youth: Youth as a Social Shifter in Botswana." *American Ethnologist* 31(4): 589–605.

Ekine, Sokari, ed. 2010. *SMS Uprising: Mobile Phone Activism in Africa*. Cape Town: Pambazuka Press.

Ellwood-Clayton, Bella. 2003. "Virtual Strangers: Young Love and Texting in the Filipino Archipelago of Cyberspace." In *Mobile Democracy*, edited by K. Nyiri, 225–35. Vienna: Passagen Verlag.

Ellwood-Clayton, Bella. 2006. "All We Need Is Love—and a Mobile Phone: Texting in the Philippines." Paper presented at Cultural Space and Public Sphere in Asia, Seoul, Korea.

Englund, Harri. 1996. "Waiting for the Portuguese: Nostalgia, Exploitation and the Meaning of Land in the Malawi-Mozambique Borderland." *Journal of Contemporary African Studies* 14(2): 157–72.

Englund, Harri. 2011. *Human Rights and African Airwaves*. Bloomington: Indiana University Press.

Evans-Pritchard, E. E. 1976. *Witchcraft, Oracles and Magic among the Azande*. Oxford: Clarendon.

Farmer, Paul. 2004. "An Anthropology of Structural Violence." *Current Anthropology* 45(3): 305–25.

Felgate, W. S. 1982. *The Tembe Thonga of Natal and Mozambique: An Ecological Approach*. Durban: University of Natal.

Ferguson, James. 1990. *The Anti-Politics Machine: "Development," Depoliticization, and Bureaucratic Power in Lesotho*. Cambridge: Cambridge University Press.

Ferguson, James. 1999. *Expectations of Modernity: Myths and Meanings of Urban Life on the Zambian Copperbelt*. Berkeley: University of California Press.

Ferguson, James. 2002. "Of Mimicry and Membership: Africans and the 'New World Society.'" *Cultural Anthropology* 17(4): 551–69.

Ferguson, James. 2006. *Global Shadows: Africa in the Neoliberal World Order*. Durham, NC: Duke University Press.

Ferguson, James. 2013. "Declarations of Dependence: Labour, Personhood, and Welfare in Southern Africa." *Journal of the Royal Anthropological Institute* 19(2): 223–42.

Ferguson, James. 2015. *Give a Man a Fish: Reflections on the New Politics of Distribution*. Durham, NC: Duke University Press.

Ferme, Mariane C. 2001. *The Underneath of Things: Violence, History and the Everyday in Sierra Leone*. Berkeley: University of California Press.

Fillip, Barbara. 2001. *Digital Divide*. Report, JICA USA. http://www.share4dev.info/telecentreskb/documents/4058.pdf.

First, Ruth. 1983. *Black Gold: The Mozambican Miner, Proletarian and Peasant*. Sussex: Harvester Press.

Fischer, Claude S. 1992. *America Calling: A Social History of the Telephone to 1940*. Berkeley: University of California Press.

Foucault, Michel. 1979. *Discipline & Punish: The Birth of the Prison*. Harmondsworth, UK: Penguin Books.

Fouquet, Thomas. 2007. "De La Prostitution Clandestine Aux Désirs De L'ailleurs: Une 'Ethnographie De L'extraversion' À Dakar." *Politique africaine* 107: 102–23.

Friedman, Jonathan. 2004. "The Political Economy of Elegance." In *Consumption and Identity*, edited by J. Friedman, 167–87. London: Routledge.

Gable, Eric. 1997. "A Secret Shared: Fieldwork and the Sinister in a West African Village." *Cultural Anthropology* 12(2): 213–33.

Gable, Eric. 2000. "The Culture Development Club: Youth, Neo-Tradition, and the Construction of Society in Guinea-Bissau." *Anthropological Quarterly* 73(4): 195–203.

Geffray, Christian. 1988. "Fragments D'un Discours Du Pouvoir (1975–1985): Du Bon Usage D'une Méconnaissance Scientifique." *Politique africaine* 29: 71–85.

Geffray, Christian. 1990. *La Cause Des Armes Au Mozambique: Anthropologie D'une Guerre Civile*. Paris: Karthala.

Geissler, P. Wissler. 2013. "Public Secrets in Public Health: Knowing Not to Know While Making Scientific Knowledge." *American Ethnologist* 40(1): 13–34.

Gell, Alfred. 1985. "How to Read a Map: Remarks on the Practical Logic of Navigation." *Man NS* 20(2): 271–86.

Gell, Alfred. (1996) 2011. "On Love." *Anthropology of this Century* 2. http://aotcpress.com/articles/love/.

Gerdes, Paulus. 2001. "Exploring Plaited Plane Patterns among the Tonga in Inhambane (Mozambique)." *Symmetry: Culture and Science* 12(1–2): 115–26.

Gershon, Ilana. 2010. *The Break Up 2.0: Disconnecting over New Media*. Ithaca: Cornell University Press.

Geschiere, Peter. 1997. *The Modernity of Witchcraft: Politics and the Occult in Postcolonial Africa*. Chicago: University of Chicago Press.

Geschiere, Peter. 2013. *Witchcraft, Intimacy, and Trust: Africa in Comparison*. Chicago: University of Chicago Press.

Giddens, Anthony. 1992. *The Transformation of Intimacy: Love, Sexuality and Eroticism in Modern Societies*. Cambridge, UK: Blackwell.

Gilsenan, Michael. 1976. "Lying, Honor, and Contradiction." In *Transaction and Meaning*, edited by B. Kapferer, 191–219. Philadelphia: Institute for the Study of Human Issues.

Ginsburg, Faye D., Lila Abu-Lughod, and Brian Larkin. 2002. "Introduction." In *Media Worlds*, edited by F. D. Ginsburg, L. Abu-Lughod, and B. Larkin, 1–36. Berkeley: University of California Press.

Gluckman, Max. 1963. "Gossip and Scandal." *Current Anthropology* 4(3): 307–16.

Gluckman, Max. 1968. "Psychological, Sociological and Anthropological Explanations of Witchcraft and Gossip: A Clarification." *Man NS* 3(1): 20–34.

Goffman, Erving. 1959. *The Presentation of Self in Everyday Life*. London: Penguin Books.

Goldstein, Daniel M. 2003. "'In Our Own Hands': Lynching, Justice, and the Law in Bolivia." *American Ethnologist* 30(1): 22–43.

Gonçalves, Euclides. 2013. "Orientações Superiores: Time and Bureaucratic Authority in Mozambique." *African Affairs* 112(449): 602–22.

Gonzalez, Olga M. 2010. *Unveiling Secrets of War in the Peruvian Andes*. Chicago: University of Chicago Press.

Goodman, Deena. 2009. *Becoming a Woman in the Age of Letters*. New York: Cornell University Press.

Greenberg, A., and G. Sadowsky. 2006. *A Country ICT Survey for Mozambique*. Montréal: Sida and Greenberg ICT Services.

Gregor, Thomas. 1977. *Mehinaku: The Drama of Daily Life in a Brazilian Indian Village*. Chicago: University of Chicago Press.

Groes-Green, Christian. 2010. "Orgies of the Moment: Bataille's Anthropology of Transgression and the Defiance of Danger in Post-Socialist Mozambique." *Critique of Anthropology* 10(4): 385–407.

Groes-Green, Christian. 2013. "'To Put Men in a Bottle': Eroticism, Kinship, Female Power, and Transactional Sex in Maputo, Mozambique." *American Ethnologist* 40(1): 102–17.

Groes-Green, Christian. 2014. "Journeys of Patronage: Moral Economies of Female Migration from Mozambique to Europe." *Journal of the Royal Anthropological Institute 20(2): 237–55.*

Habermas, J. 1989. *The Structural Transformation of the Public Sphere: An Inquiry into a Category of Bourgeois Society*. Cambridge, UK: Polity.

Hahn, Hans Peter. 2012. "Mobile Phones and the Transformation of Society: Talking about Criminality and the Ambivalent Perception of New ICT in Burkina Faso." *African Indentities* 10(2): 181–92.

Hahn, Hans Peter, and Ludovic Kibora. 2008. "The Domestication of the Mobile Phone: Oral Society and New ICT in Burkina Faso." *Journal of Modern African Studies* 46(1): 87–109.

Hanlon, Joseph. 1996. *Peace without Profit: How the IMF Blocks Rebuilding in Mozambique*. Oxford: James Currey.

Hanlon, Joseph. 2007. "Is Poverty Decreasing in Mozambique?" Paper presented at Inaugural Conference of the Institute of Social and Economic Studies, Maputo.

Hansen, Karen T. 2005. "Getting Stuck in the Compound: Some Odds against Social Adulthood in Lusaka, Zambia." *Africa Today* 51(4): 3–16.

Hansen, Thomas Blom, and Oska Verkaaik. 2009. "Urban Charisma: On Everyday Mythologies in the City." *Critique of Anthropology* 29(1): 5–26.

Haram, Liv. 2005. "'Eyes Have No Curtains': The Moral Economy of Secrecy in Managing Love Affairs among Adolescents in Northern Tanzania in the Time of Aids." *Africa Today* 51(4): 57–73.

Harries, Patrick. 1994. *Work, Culture and Identity*. Portsmouth: Heinemann.

Harris, Marvin. 1959. "Labour Emigration among the Mozambique Thonga: Cultural and Political Factors." *Africa* 29: 50–66.

Hart, Keith. 2010. "Informal Economy." In *The Human Economy: A Citizen's Guide*, edited by K. Hart, J. Laville, and A. D. Cattani, 142–54. Cambridge, UK: Polity.

Hawkins, K., F. Mussá, and S. Abuxahama. 2005. *"Milking the Cow": Young Women's Constructions of Identity, Gender, Power and Risk in Transactional and Cross-Generational Sexual Relationships: Maputo, Mozambique*. Maputo: Population Services International and USAID.

Heald, Suzete. 1995. "The Power of Sex: Some Reflections on the Caldwells' 'African Sexuality' Thesis." *Africa* 65(4): 489–505.

Heald, Suzette. 1999. *Manhood and Morality: Sex, Violence and Ritual in Gisu Society*. London: Routledge.

Helgesson, Alf. 1994. *Church, State and People in Mozambique*. PhD diss., Uppsala University.

Herzfeld, Michael. 2004. "Intimating Culture: Local Contexts and International Power." In *Off Stage/On Display: Intimacy and Ethnography in the Age of Public Culture*, edited by A. Shryock, 317–35. Stanford, CA: Stanford University Press.

Hirsch, Jennifer, Holly Wardlow, and Harriet Phinney. 2012. "'No One Saw Us': Reputation as an Axis of Sexual Identity." In *Understanding Global Sexualities: New Frontiers*, edited by P. Aggleton, P. Boyce, H. Moore, and R. Parker, 91–107. New York: Routledge.

Hirsch, Jennifer S., Holly Wardlow, Daniel Jordan Smith, Harriet M. Phinney, Shanti Parikh, and Constance A. Nathanson. 2010. *The Secret: Love, Marriage, and HIV*. Nashville: Vanderbilt University Press.

Holbraad, Martin. 2012. *Truth in Motion: The Recursive Anthropology of Cuban Divination*. Chicago: University of Chicago Press.

Honwana, Alcinda. 2003. "Undying Past: Spirit Possession and the Memory of War in Southern Mozambique." In *Magic and Modernity*, edited by B. Meyer and P. Pels, 60–80. Stanford, CA: Stanford University Press.

Honwana, Alcinda. 2005. "Innocent and Guilty: Child-Soldiers as Interstitial and Tactical Agents." In *Makers and Breakers: Children & Youth in Postcolonial Africa*, edited by A. Honwana and F. De Boeck, 31–52. Oxford: James Currey.

Honwana, Alcinda. 2012. *The Time of Youth: Work, Social Change, and Politics in Africa*. Washington: Kumarian Press.

Horst, Heather. 2009. "Aesthetics of the Self: Digital Mediations." In *Anthropology and the Individual: A Material Culture Perspective*, edited by D. Miller, 99–113. Oxford: Berg.

Horst, Heather A., and Daniel Miller. 2006. *The Cell Phone: An Anthropology of Communication*. Oxford: Berg.

Hull, Elizabeth, and Deborah James. 2012. "Introduction: Popular Economies in South Africa." *Africa* 82(1): 1–19.

Humphrey, Caroline. 1999. "Russian Protection Rackets and the Appropriation of Law and Order." In *States and Illegal Practices*, edited by J. Heyman, 199–232. Oxford: Berg.

Hunter, Mark. 2002. "The Materiality of Everyday Sex: Thinking Beyond 'Prostitution.'" *African Studies* 61(1): 99–120.

Hunter, Mark. 2009. "Providing Love: Sex and Exchange in Twentieth-Century South Africa." In *Love in Africa*, edited by J. Cole and L. M. Thomas, 135–56. Chicago: University of Chicago Press.

Hunter, Mark. 2010. *Love in the Time of AIDS: Inequality, Gender, and Rights in South Africa*. Bloomington: Indiana University Press.

Isaacman, Allen. 1978. *A Luta Continua: Creating a New Society in Mozambique*. Binghamton: State University of New York.

Ito, Mizuko, Daisuke Okabe, and Misa Matsuda, eds. 2005. *Personal, Portable, Pedestrian*. Cambridge, MA: MIT Press.

Ivaska, Andrew. 2011. *Cultured States: Youth, Gender, and Modern Style in 1960s Dar Es Salaam*. Durham, NC: Duke University Press.

Jackson, Michael. 1998. *Minima Ethnographica: Intersubjectivity and the Anthropological Project*. Chicago: University of Chicago Press.

Jackson, Michael. 2013. *Lifeworlds: Essays in Existential Anthropology*. Chicago: University of Chicago Press.

James, Deborah. 2011. "The Return of the Broker: Consensus, Hierarchy, and Choice in South African Land Reform." *Journal of the Royal Anthropological Institute* 17: 318–38.

James, Deborah. 2012. "Money-Go-Round: Personal Economies of Wealth, Aspiration and Indebtedness." *Africa* 82(1): 20–40.

Jeffrey, Craig. 2010. *Timepass: Youth, Class and the Politics of Waiting*. Stanford, CA: Stanford University Press.

Jeffrey, Craig, and Jane Dyson. 2013. "Zigzag Capitalism: Youth Entrepreneurship in the Contemporary Global South." *Geoforum* 49: R1–R3.

Jensen, Robert. 2007. "The Digital Provide: Information (Technology), Market Performance, and Welfare in the South Indian Fisheries Sector." *Quarterly Journal of Economics* CXXII(3): 879–924.

Jensen, Steffen. 2008. "Policing Nkomazi: Crime, Masculinity and Generational Conflicts." In *Global Vigilantes*, edited by D. Pratten and A. Sen, 47–68. New York: Columbia University Press.

Johnson-Hanks, Jennifer. 2002. "On the Limits of Life-Story in Ethnography: Towards a Theory of Vital Conjunctures." *American Anthropologist* 1004(3): 865–80.

Johnson-Hanks, Jennifer. 2005. "When the Future Decides: Uncertainty and Intentional Action in Contemporary Cameroon." *Current Anthropology* 46(3): 363–85.

Jones, Graham M. 2014. "Secrecy." *Annual Review of Anthropology* 43: 53–69.

Jones, Jeremy L. 2010. "'Nothing Is Straight in Zaimbabwe': The Rise of the Kuyika-Kiya Economy, 2000–2008." *Journal of Southern African Studies* 36(2): 285–99.

Junod, Henri A. (1912) 1966. *The Life of a South African Tribe*. New York: University Books.

Katz, James. E., and M. Aakhus, eds. 2002. *Perpetual Contact: Mobile Communication Private Talk, Public Performance*. Cambridge: Cambridge University Press.

Krige, Eileen Jensen, and John L. Comaroff, eds. 1981. *Essays on African Marriage in Southern Africa*. Cape Town: Ruta.

Kuper, Adam. 1982. *Wives for Cattle: Bridewealth and Marriage in Southern Africa*. London: Routledge Kegan & Paul.

Kyed, Helen Maria. 2008. "State Vigilantes and Political Community on the Margins in Post-War Mozambique." In *Global Vigilantes*, edited by D. Pratten and A. Sen, 393–415. New York: Columbia University Press.

Lambek, Michael, ed. 2010. *Ordinary Ethics: Anthropology, Language, and Action*. New York: Fordham University Press.

Lamont, Tom. 2016. "End of the Affairs." *Observer Magazine* (February 28, 2016), 18–27.

Langevang, T., and K. V. Gough. 2009. "Surviving through Movement: The Mobility of Urban Youth in Ghana." *Social & Cultural Geography* 10(7): 741–56.

Larkin, Brian. 2004. "Degraded Images, Distorted Sounds: Nigerian Video and the Infrastructure of Piracy." *Public Culture* 16(2): 289–314.

Larkin, Brian. 2008. *Signal and Noise: Media, Infrastructure, and Urban Culture in Nigeria*. Durham, NC: Duke University Press.

Larkin, Brian. 2013. "The Politics and Poetics of Infrastructure." *Annual Review of Anthropology* 42: 327–43.

Lin, Angel M. Y., and Avin H. M. Tong. 2007. "Text-Messaging Cultures of College Girls in Hong Kong: SMS as Resources for Achieving Intimacy and Gift-Exchange with Multiple Functions." *Continuum: Journal of Media and Cultural Studies* 21(2): 303–15.

Ling, Rich. 2008. *New Tech, New Ties*. Cambridge, MA: MIT Press.

Ling, Rich. 2004. *The Mobile Connection: The Cell Phone's Impact on Society*. San Francisco: Morgan Kaufmann.

Ling, Rich, and Per E. Pedersen, eds. 2005. *Mobile Communications: Re-Negotiations of the Social Sphere*. London: Springer.

Locoh, Thérèse. 1994. "Social Change and Marriage Arrangements: New Types of Unions in Lomé, Togo." In *Nuptiality in Sub-Saharan Africa*, edited by C. Bledsoe and G. Pison, 215–30. Oxford: Clarendon Press.

Lubkemann, Stephen. 2007. *Culture in Chaos: An Anthropology of the Social Condition in War*. Chicago: University of Chicago Press.

Lubkemann, Stephen, and D. Hoffman. 2005. "Warscape Ethnography in West Africa and the Anthropology of 'Events.'" *Anthropological Quarterly* 78(2): 315–28.

Lukacs, Gabriella. 2013. "Dreamwork: Cell Phone Novelists, Labor, and Politics in Contemporary Japan." *Cultural Anthropology* 28(1): 44–64.

Lukose, Ritty A. 2012. "'Shining' in Public: Masculine Assertion and Anxiety in Globalizing Kerala." In *Young Men in Uncertain Times*, edited by V. Amit and N. Dyck, 35–58. New York: Berghahn Books.

Macamo, Elísio. 2005. "Denying Modernity: The Regulating of Native Labour in Colonial Mozambique and Its Postcolonial Aftermath." In *Negotiating Modernity: Africa's Ambivalent Experience*, edited by E. Macamo, 67–97. Dakar: Codesria Books.

Macamo Raimundo, Inês. 2005. "From Civil War to Floods: Implications for Internal Migration in Gaza Province of Mozambique." In *Negotiating Modernity: Africa's Ambivalent Experience*, edited by E. Macamo, 159–71. Dakar: Codesria Books.

Machava, Benedito Luís. 2011. "State Discourse on Internal Security and the Politics of Punishment in Post-Independence Mozambique (1975–1983)." *Journal of Southern African Studies* 37(3): 593–609.

Mains, Daniel. 2007. "Neoliberal Times: Progress, Boredom, and Shame among Young Men in Urban Ethiopia." *American Ethnologist* 34(4): 659–73.

Mains, Daniel. 2012. *Hope Is Cut: Youth, Unemployment, and the Future in Urban Ethiopia*. Philadelphia: Temple University Press.

Mair, Jonathan, Ann H. Kelly, and Cassey High. 2012. "Introduction: Making Ignorance an Ethnographic Object." In *The Anthropology of Ignorance: An Ethnographic Approach*, edited by C. High, A. H. Kelly, and J. Mair. New York: Palgrave Macmillan.

Malinowski, Bronislav. (1926) 1966. *Crime and Custom in Savage Society*. London: Routledge and Kegan Paul.

Mantz, Jeffrey. 2008. "Improvisational Economies: Coltan Production in the Eastern Congo." *Social Anthropology* 16(1): 34–50.

Manuel, Sandra. 2008. *Love and Desire: Concepts, Narratives and Practices of Sex Amongst Youths in Maputo*. Dakar: Codesria.

Marchand, Trevor. 2009. *The Masons of Djenné*. Bloomington: Indiana University Press.

Mário, Mouzinho, Peter Fry, Lisbeth Levey, and Arlindo Chilundo. 2003. *Higher Education in Mozambique*. Oxford: James Currey.

Maroon, Bahiyyih. 2006. "Mobile Sociality in Urban Morocco." In *The Cell Phone Reader: Essays in Social Transformation*, edited by A. Kavoori and N. Arceneaux, 189–204. New York: Peter Lang.

Marshall, Judith. 1993. *Literacy, Power, and Democracy in Mozambique: The Governance of Learning from Colonization to the Present*. Boulder, CO: Westview Press.

Masquelier, Adeline. 2002. "Road Mythograohies: Space, Mobility, and the Historical Imagination in Postcolonial Niger." *American Ethnologist* 29(4): 829–56.

Masquelier, Adeline. 2005. "The Scorpion's Sting: Youth, Marriage and the Struggle for Social Maturity in Niger." *Journal of the Royal Anthropological Institute* 11: 59–83.

Masquelier, Adeline. 2013. "Teatime: Boredom and the Temporalities of Young Men in Niger." *Africa* 83(3): 470–91.

Mate, Rekopantswe. 2002. "Wombs as God's Laboratories: Pentecostal Discourses of Femininity in Zimbabwe." *Africa* 72(4): 549–68.

Mazrui, Ali A., and Charles C. Okigbo. 2004. "The Triple Heritage: The Split Soul of a Continent." In *Development and Communication in Africa*, edited by C. C. Okigbo and F. Eribo, 15–30. Lanham: Rowman and Littlefield Publishers.

Mazzarella, Willam. 2006. "Internet X-Ray: E-Governance, Transparency, and the Politics of Immediation in India." *Public Culture* 18: 3.

Mazzarella, William. 2010. "Beautiful Balloon: The Digital Divide and the Charisma of New Media in India." *American Ethnologist* 37(4): 783–804.

Mbembe, Achilles. 1988. *Afriques Indociles. Christianisme, Pouvoir Et État En Société Postcoloniale*. Paris: Karthala.

Mbembe, Achilles. 2000. "À Propos Des Écritures Africaines De Soi." *Politique africaine* 77: 16–43.

Mbembe, Achilles, and Janet Roitman. 1995. "Figures of the Subject in Times of Crisis." *Public Culture* 7: 323–52.

McIntosh, Janet. 2010. "Mobile Phones and Mipoho's Prophecy: The Powers and Dangers of Flying Language." *American Ethnologist* 37(2): 337–53.

Melkote, Srinivas R., and H. Leslie Steeves. 2004. "Information and Communication Technologies for Rural Development." In *Development and Communication in Africa*, edited by C. C. Okigbo and F. Eribo, 165–73. Oxford: Rowman & Littlefield.

Mercer, Claire. 2004. "Engineering Civil Society: ICT in Tanzania." *Review of African Political Economy* 99: 49–64.

Miller, Daniel, and Heather A. Horst. 2012. "The Digital and the Human: A Prospectus for Digital Anthropology." In *Digital Anthropology*, edited by H. A. Horts and D. Miller, 3–35. London: Berg.

Miller, Daniel, Andrew Skuse, Don Slater, Jo Tacchi, Tripta Chandola, Thomas Cousins, Heather Horst, and Janet Kwami. 2005. *Information Society: Emergent Technologies and Development Communities in the South*. Final Report, Information Society Research Group. http://www.share4dev.info/kb/documents/3274.pdf.

Miller, Daniel, and Don Slater. 2000. *The Internet*. New York: Berg.

Molony, Thomas. 2008. "Running out of Credit: The Limitations of Mobile Telephony in a Tanzanian Agricultural Marketing System." *Journal of Modern African Studies* 46(4): 637–58.

Morier-Genoud, Eric. 1996. "Of God and Caesar: The Relation between Christian Churches and the State in Post-Colonial Mozambique, 1974–1981." *Le Fait Missionnaire*, Cahier no. 3.

Muchie, Mammo, and Angathevar Baskaran. 2006. "Introduction." In *Bridging the Digital Divide: Innovation Systems for ICT in Brazil, China, India, Thailand and Southern Africa*, edited by A. Baskaran and M. Muchie, 23–50. London: Adonis & Abbey Publishers Ltd.

Murphy, William P. 1980. "Secret Knowledge as Property and Power in Kpelle Society: Elders versus Youth." *Africa* 50(2): 193–207.

Myerson, G. 2001. *Heidegger, Habermas and the Mobile Phone*. London: Icon Books.

Newell, Sasha. 2012. *The Modernity Bluff: Crime, Consumption, and Citizenship in Côte D'ivoire*. Chicago: University of Chicago Press.

Nielinger, Olaf. 2006. *Information and Communication Technologies (ICT) for Development in Africa*. Frankfurt: Peter Lang.

Nyamnjoh, Francis B. 2008. *Married but Available*. East Lansing: Michigan State University Press.

Nyanzi, S., J. Nassimbwa, V. Kayizzi, and S. Kabanda. 2008. "'African Sex Is Dangerous!' Renegotiating 'Ritual Sex' in Contemporary Masaka District." *Africa* 78(4): 518–39.

Obarrio, Juan M. 2010. "Remains: To Be Seen. Third Encounter between State and 'Customary' in Northern Mozambique." *Cultural Anthropology* 25(2): 263–300.

O'Laughlin, Bridget. 1995. "Past and Present Options: Land Reform in Mozambique." *Review of African Political Economy* 22(63): 99–106.

O'Laughlin, Bridget. 2000. "Class and the Customary: The Ambiguous Legacy of the *Indigenato* in Mozambique." *African Affairs* 99: 5–42.

Ong, Aihwa, and Stephen Collier, eds. 2005. *Global Assemblages: Technology, Politics, and Ethics as Anthropological Problems*. Oxford: Blackwell.

Orlove, Benjamin. 2005. "Editorial: Time, Society, and the Course of New Technologies." *Current Anthropology* 46(5): 699–700.

Ortner, Sherry. 2006. *Anthropology and Social Theory: Culture, Power, and the Acting Subject*. Durham, NC: Duke University Press.

Osborn, Michelle. 2008. "Fuelling the Flames: Rumour and Politics in Kibera." *Journal of Eastern African Studies* 2(2): 315–27.

Paine, Robert. 1967. "What Is Gossip About? An Alternative Hypothesis." *Man (N. S.)* 2(2): 278–85.

Parry, Jonathan, and Maurice Bloch. 1989. "Introduction: Money and the Morality of Exchange." In *Money and the Morality of Exchange*, edited by J. Parry and M. Bloch, 1–32. Cambridge: Cambridge University Press.

Pélissier, René. 1984. *Naissance Du Mozambique. Résistance Et Révoltes Anticoloniales (1854–1918)*. Orgeval, Yvelines: Pélissier.

Pelkmans, Lotte. 2009. "Phoning Anthropologists: The Mobile Phone's (Re)Shaping of Anthropological Research." In *The New Talking Drum*, edited by M. de Bruijn, F. Nyamnjoh, and I. Brinkman, 23–49. Leiden: Langaa & African Studies Centre.

Penvenne, Jeanne. 1995. *African Workers and Colonial Racism*. London: James Currey.

Peters, Krijn. 2011. *War and the Crisis of Youth in Sierra Leone*. Cambridge: Cambridge University Press.

Pfeiffer, James. 2004. "Condom Social Marketing, Pentecostalism, and Structural Adjustment in Mozambique: A Clash of Aids Prevention Messages." *Medical Anthropology Quarterly* 18(1): 77–103.

Pfeiffer, James. 2006. "Money, Modernity, and Morality: Traditional Healing and the Expansion of the Holy Spirit in Mozambique." In *Borders and Healers: Brokering Therapeutic Resources in Southeast Africa*, edited by T. J. Luedke and H. G. West, 81–100. Bloomington: Chesham Indiana University Press.

Pfeiffer, James, Kenneth Sherr-Gimbel, and Orvalho Joaquim Augusto. 2007. "The Holy Spirit in the Household: Pentecostalism, Gender, and Neoliberalism in Mozambique." *American Anthropologist* 109(4): 688–700.

Piot, Charles. 1993. "Secrecy, Ambiguity, and the Everyday in Kabre Culture." *American Anthropologist* 95(2): 353–70.

Piot, Charles. 2010. *Nostalgia for the Future: West Africa after the Cold War*. Chicago: University of Chicago Press.

Pitcher, M. Anne. 2006. "Forgetting from Above and Memory from Below: Strategies of Legitimation and Struggle in Postsocialist Mozambique." *Africa* 76(1): 88–112.

Pitcher, M. Anne, and Kelly M. Askew. 2006. "African Socialism and Postsocialisms." *Africa* 76(1): 1–14.

Plant, Sadie. 2001. *On the Mobile: The Effects of Mobile Telephones on Social and Individual Life*. Motorola. http://www.momentarium.org/experiments/7a10me/sadie_plant.pdf.

Porter, G., K. Hampshire, A. Abane, E. Robson, A. Munthali, M. Mashiri, and A. Tanle. 2010. "Moving Young Lives: Mobility, Immobility and Inter-Generational Tensions in Urban Africa." *Geoforum* 41(5): 796–804.

Povinelli, Elizabeth. 2006. *The Empire of Love: Towards a Theory of Intimacy, Genealogy, and Carnality*. Durham, NC: Duke University Press.

Pratten, David. 2008a. "Masking Youth: Transformation and Transgression in Annang Performance." *African Arts* 41(4): 44–60.

Pratten, David. 2008b. "Singing Thieves: History and Practice in Nigerian Social Justice." In *Global Vigilantes*, edited by D. Pratten and A. Sen, 175–205. New York: Columbia University Press.

Prince, Ruth. 2006. "Popular Music and Luo Youth in Western Kenya: Ambiguities of Modernity, Morality and Gender Relations in the Era of Aids." In *Navigating Youth, Generating Adulthoods: Social Becoming in an African Context*, edited by C. Christiansen, M. Utas, and H. E. Vigh, 117–52. Stockholm: Nordic Africa Institute.

Raffles, Hugh. 2002. "Intimate Knowledge." *International Social Science Journal* 54(173): 325–35.

Rasmussen, Susan J. 2000. "Between Several Worlds: Images of Youth and Age in Tuareg Popular Performances." *Anthropological Quarterly* 73(3): 133–44.

Reed, Adam. 1999. "Anticipating Individuals: Modes of Vision and Their Social Consequence in a Papua New Guinea Prison." *Journal of the Royal Anthropological Institute* 5(1): 46–56.

Rhine, Kathryn. 2014. "HIV, Embodied Secrets, and Intimate Labour in Northern Nigeria." *Ethnos* 79(5): 699–718.

Robins, Kevin, and Frank Webster. 1999. *Times of the Technoculture*. London: Routledge.

Roesch, Otto. 1992. "Renamo and the Peasantry in Southern Mozambique: A View from Gaza Province." *Canadian Journal of African Studies* 26: 462–84.

Roitman, Janet. 2005. "The Garrison-Entrepôt: A Mode of Governing in the Chad Basin." In *Global Assemblages: Technology, Politics, and Ethics as Anthropological Problems*, edited by A. Ong and S. J. Collier, 417–36. Malden, MA: Blackwell.

Roitman, Janet. 2014. *Anti-Crisis*. Durham, NC: Duke University Press.

Rungo, Jorge. 2007. "População mostra maturidade e as autoridades vão a reboque." *Domingo*, Maputo no. 1329: 4–5.

Sahlins, Marshall. 1993. "Goodbye to *Tristes Tropes*: Ethnography in the Context of Modern World History." *Journal of Modern History* 65: 1–25.

Sanders, Todd. 2001. "Save Our Skins: Structural Adjustment, Morality and the Occult in Tanzania." In *Magical Interpretations, Material Realities*, edited by H. L. Moore and T. Sanders, 160–83. London: Routledge.

Sarró, Ramon. 2009. *The Politics of Religious Change on the Upper Guinea Coast: Iconoclasm Done and Undone*. Edinburg: Edinburg University Press.

Sartre, Jean-Paul. 1943. *L'être Et Le Néant*. Paris: Gallimard.

Sebastião, Lica. 1999. "Hoje não existem moças para casar." *Português 11a Classe. Textos E Sugestões De Actvidades*. Lisbon: Diname. 50.

Schapera, Isaac. 1941. *Married Life in an African Tribe*. New York: Sheridan House.

Schneider, Jane, and Peter Schneider. 2008. "The Anthropology of Crime and Criminalization." *Annual Review of Anthropology* 37: 351–73.

Schroeder, Richard A. 1996. " 'Gone to Their Second Husbands': Marital Metaphors and Conjugal Contracts in the Gambia's Female Garden Sector." *Canadian Journal of African Studies* 30(1): 69–87.

Sciriha, Lydia. 2006. "Teenagers and Mobiles Phones in Malta: A Sociolinguistic Profile." In *New Technologies in Global Societies*, edited by P.-L. Law, L. Fortunati, and S. Yang, 159–78. Singapore: World Scientific Publishing Co.

Serra, Carlos. 2008. Linchamentos, Eclipse Do Social E Bodes Expiatórios. Accessed March 4, 2012. http://oficinadesociologia.blogspot.co.uk/2008/03/linchamentos-eclipse-do-social-e-bodes.html.

Sey, Araba. 2011. "'We Use It Different, Different': Making Sense of Trends in Mobile Phone Use in Ghana." *New Media and Society* 13(3): 375–90.

Sharma, Nitasha Tamar. 2010. *Hip Hop Desis: South Asian Americans, Blackness, and a Global Race Consciousness*. Durham, NC: Duke University Press.

Shaw, Rosalind. 2002. *Memories of the Slave Trade: Ritual and the Social Imagination in Sierra Leone*. Chicago: University of Chicago Press.

Sheldon, Kathleen. 1998. "'I Studied with the Nuns, Learning to Make Blouses': Gender Ideology and Colonial Education in Mozambique." *International Journal of African Historical Studies* 31(3): 595–625.

Sheldon, Kathleen E. 2002. *Pounders of Grain: A History of Women, Work, and Politics in Mozambique*. Portsmouth, NH: Heinemann.

Shipton, Parker. 2007. *The Nature of Entrustment: Intimacy, Exchange and the Sacred in Africa*. New Haven: Yale University Press.

Shryock, Andrew. 2004. "Other Conscious/Self Aware: First Thoughts on Cultural Intimacy and Mass Mediation." In *Off Stage/On Display: Intimacy and Ethnography in the Age of Public Culture*, edited by A. Shryock, 3–28. Stanford, CA: Stanford University Press.

Silberschmidt, Margrethe. 2004. "Masculinities, Sexuality and Socio-Economic Change in Rural and Urban East-Africa." In *Re-Thinking Sexualities in Africa*, edited by S. Arnfred, 233–48. Stockholm: Almqvist & Wiksell Tryckeri.

Silberschmidt, Margrethe. 2005. "Poverty, Male Disempowerment, and Male Sexuality: Rethinking Men and Masculinities in Rural and Urban East Africa." In *African Masculinities: Men in Africa from the Late Nineteenth Century to the Present*, edited by L. Ouzgane and R. Morrell, 189–203. New York: Palgrave Macmillan.

Silverstone, Roger, Eric Hirsch, and David Morley. 1992. "Information and Communication Technologies and the Moral Economy of the Household." In *Consuming Technologies: Media and Information in Domestic Space*, edited by R. Silverstone and E. Hirsch, 15–31. London: Routledge.

Simmel, Georg. 1906. "The Sociology of Secrecy and of Secret Societies." *American Journal of Sociology* 11(4): 441–98.

Simmel, Georg. 1950. *The Sociology of Georg Simmel*. London: Collier Macmillan.

Simmel, Georg. (1900) 1978. *The Philosophy of Money*. London: Routledge and Kegan Paul.

Simone, AbdouMaliq. 2004. "People as Infrastructure: Intersecting Fragments in Johannesburg." *Public Culture* 16(3): 407–29.

Simone, AbdouMaliq. 2006. "Pirate Towns: Reworking Social and Symbolic Infrastructures in Johannesburg and Douala." *Urban Studies* 43(2): 357–70.

Skuse, Andrew, and Thomas Cousins. 2007. "Managing Distance: Rural Poverty and the Promise of Communication in Post-Apartheid South Africa." *Journal of Asian and African Studies* 42(2): 185–207.

Slater, Don, and Janet Kwami. 2005. "Embeddedness and Escape: Internet and Mobile Use as Poverty Reduction Strategies in Ghana." Accessed September, 30, 2009, http://zunia.org/uploads/media/knowledge/internet.pdf.

Smith, Alan K. 1973. "The Peoples of Southern Mozambique: An Historical Survey." *Journal of African History* XIV(4): 565–80.

Smith, Daniel Jordan. 2006. "Cell Phones, Social Inequality, and Contemporary Culture in Nigeria." *Canadian Journal of African Studies* 40(3): 496–523.

Smith, David. 2012. "Boom Time for Mozambique, Once the Basket Case of Africa." *The Guardian*, March 27, https://www.theguardian.com/world/2012/mar/27/mozambique-africa-energy-resources-bonanza.

Smith, James H. 2011. "Tantalus in the Digital Age: Coltan Ore, Temporal Dispossession, and 'Movement' in Eastern Democratic Republic of the Congo." *American Ethnologist* 38(1): 17–35.

Sommers, Marc. 2012. *Stuck: Rwandan Youth and the Struggle for Adulthood*. Athens: University of Georgia Press.

Spencer, Paul. 1998. *The Pastoral Continuum: The Marginalization of Tradition in East Africa*. Oxford: Clarendon Press.

Spronk, Rachel. 2009. "Sex, Sexuality and Negotiating Africanness in Nairobi." *Africa* 79(4): 500–519.

Stambach, Amy, and George A. Malekela. 2006. "Education, Technology, and the 'New' Knowledge Economy: Views from Bongoland." *Globalisation, Societies and Education* 4(3): 321–36.

Stark, Laura. 2013. "Transactional Sex and Mobile Phones in a Tanzania Slum." *Journal of the Finish Anthropological Society* 38(1): 12–36.

Steffen, Vibeke, Richard Jenkins, and Hanne Jessen, eds. 2005. *Managing Uncertainty: Ethnographic Studies of Illness, Risk and the Struggle for Control*. Copenhagen: Museum Tusculanum Press.

Steinberg, Jonny. 2014. *A Man of Good Hope*. Denver, South Africa: Jonathan Ball.

Stephens, Sharon. 1995. "Children and the Politics of Culture in 'Late Capitalism.'" In *Children and the Politics of Culture*, edited by S. Stephens, 3–48. Princeton, NJ: Princeton University Press.

Sumich, Jason. 2008. "Politics after the Time of Hunger in Mozambique: A Critique of Neo-Patrimonial Interpretation of African Elites." *Journal of Southern African Studies* 34(1): 111–25.

Sunderland, P. L. 1999. "Fieldwork and the Phone." *Anthropological Quarterly* 72(3): 105–17.

Taussig, M. 1987. *Shamanism, Colonialism, and the Wild Man: A Study in Terror and Healing*. Chicago: University of Chicago Press.

Taussig, Michael. 1999. *Defacement: Public Secrecy and the Labor of the Negative*. Stanford, CA: Stanford University Press.

Taussig, Michael. 2012. *Beauty and the Beast*. Chicago: University of Chicago Press.

Tenhunen, Sirpa. 2008. "Mobile Technology in the Village: ICT, Culture, and Social Logistics in India." *Journal of the Royal Anthropological Institute* 14: 515–34.

Thioune, Ramata Molo, ed. 2003. *Information and Communication Technologies for Development in Africa*. Ottawa: International Development Research Centre.

Thomas, Lynn M. 2006. "Schoolgirl Pregnancies, Letter-Writing, and 'Modern' Personhood Persons in Late Colonial East Africa." In *Africa's Hidden Histories: Everyday Literacy and Making the Self*, edited by K. Barber, 180–207. Bloomington: Indiana University Press.

Thomas, Lynn M., and Jennifer Cole. 2009. "Thinking through Love in Africa." In *Love in Africa*, edited by J. Cole and L. M. Thomas, 1–30. Chicago: University of Chicago Press.

Thompson, Mark. 2004. "Discourse, 'Development' and the 'Digital Divide': ICT and the World Bank." *Review of African Political Economy* 99: 103–23.

Thornton, Robert. 2008. *Unimagined Community: Sex, Networks, and AIDS in Uganda and South Africa*. Berkeley: University of California Press.

Trawick, Margaret. 1990. *Notes on Love in a Tamil Family*. Berkeley: University of California Press.

Traxler, John, and Philip Dearden. 2005. "The Potential for Using SMS to Support Learning and Organisation in Sub-Saharan Africa." http://www2.wlv.ac.uk/webteam/about/cidt/DSA%20 Submission.pdf.

Trottier, Daniel. 2012. *Social Media as Surveillance: Rethinking Visibility in a Converging World*. Farnham, UK: Ashgate.

Tsing, Anna. 2000. "Inside the Economy of Appearances." *Public Culture* 12(1): 115–44.

Vail, Leroy, and Landeg White. 1991. *Power and the Praise Poem*. London: James Curry.

van Binsbergen, Wim. 2004. "Can ICT Belong in Africa, or Is ICT Owned by the North Atlantic Region?" In *Situating Globality: African Agency in the Appropriation of Global Culture*, edited by W. van Binsbergen and R. van Dijk, 107–55. Leiden: Brill.

van de Kamp, Linda. 2012. "Love Therapy: A Brazilian Pentecostal (Dis)Connection in Maputo." In *The Social Life of Connectivity in Africa*, edited by M. de Bruijn and R. van Dijk, 203–25. New York: Palgrave Macmillan.

van der Drift, Roy. 2002. "Democracy's Heady Brew: Cashew Wine and the Authority of Elders among the Balanta in Guinea-Bissau." In *Alcohol in Africa: Mixing Business, Pleasure and Politics*, edited by D. F. Bryceson, 179–96. Portsmouth, NH: Heinemann.

van Dijk, Rijk. 2012. "A Ritual Connection: Urban Youth Marrying in the Village in Botswana." In *The Social Life of Connectivity in Africa*, edited by M. de Bruijn and R. van Dijk, 141–59. New York: Palgrave Macmillan.

van Vleet, Krista. 2003. "Partial Theories: On Gossip, Envy and Ethnography in the Andes." *Ethnography* 4(4): 491–519.

Vasquez-Leon, Marcela. 1999. "Neoliberalism, Environmentalism, and Scientific Knowledge: Redefining Use Rights in the Gulf of California Fisheries." In *States and Illegal Practices*, edited by J. M. Heyman, 233–60. Oxford: Berg.

Vigh, Henrik. 2006. *Navigating Terrains of War: Youth and Soldiering in Guinea-Bissau*. New York: Berghahn Books.

Vincent, Jane. 2005. "Emotional Attachment to Mobile Phones: An Extraordinary Relationship." In *Mobile World: Past, Present and Future*, edited by L. Hamill and A. Lasen. New York: Springer.

Vines, Alex. 1991. *Renamo Terrorism in Mozambique*. York: Centre for Southern African Studies, University of York.

Wacquant, Loïc. 2004. *Body and Soul: Ethnographic Notebooks of an Apprentice-Boxer*. New York: Oxford University Press.

Webster, David. 1975. *Kinship and Co-Operation: Agnation, Alternative Structures and the Individual in Chopi Society*. PhD diss., Rhodes University.

Weiss, Brad. 2002. "Thug Realism: Inhabiting Fantasy in Urban Tanzania." *Cultural Anthropology* 17(1): 93–124.

Weiss, Brad. 2005. "The Barber in Pain: Consciousness, Affliction and Alterity in East Africa." In *Makers and Breakers: Children and Youth in Postcolonial Africa*, edited by F. De Boeck and A. Honwana, 102–20. Oxford: James Currey.

Weiss, Brad. 2009. *Street Dreams and Hip Hop Barbershops: Global Fantasy in Urban Tanzania*. Bloomington: Indiana University Press.

West, Harry G. 2005. *Kupilikula: Governance and the Invisible Realm in Mozambique*. Chicago: University of Chicago Press.

West, Harry G. 2007. *Ethnographic Sorcery*. Chicago: University of Chicago Press.

West, Harry G., and Jo Hellen Fair. 1993. "Development Communication and Popular Resistance in Africa: An Examination of the Struggle over Tradition and Modernity." *African Studies Review* 36(1): 91–114.

Whyte, Susan Reynolds. 1997. *Questioning Misfortune*. Cambridge: Cambridge University Press.

Whyte, Susan Reynolds. 2009. "Epilogue." In *Dealing with Uncertainty in Contemporary African Lives*, edited by L. Haram and B. Yamba, 213–16. Uppsala, Sweden: Nordiska Afrikainstitutet.

Williams, Raymond. 1974. *Television: Technology and Cultural Form*. Glasgow: Fontana.

Willis, Justin. 2002. *Potent Brews: A Social History of Alcohol in East Africa, 1850–1999*. Athens: Ohio University Press.

Wilson, Ken. 1992. "Cults of Violence and Counter-Violence in Mozambique." *Journal of Southern African Studies* 18(3): 527–82.

Wittgenstein, Ludwig. (1969) 1975. *On Certainty*. Oxford: Blackwell.

Wuyts, Marc Eric. 1989. *Money and Planning for Socialist Transition: The Mozambican Experience*. Aldershot, UK: Gower.

Zigon, Jarrett. 2009. "Hope Dies Last: Two Aspects of Hope in Contemporary Moscow." *Anthropological Theory* 9(3): 253–71.

Index